Other titles of interest

Seven Years of MANIFOLD: 1968–1980
Ian Stewart & John Jaworski (Editors)

The Collected Letters of Colin MacLaurin
Stella Mills (Editor)

In the Shiva Mathematics Series

Categories, Bundles and Spacetime Topology
C.T.J. Dodson

Group Tables
A.D. Thomas & G.V. Wood

Manifolds of Differentiable Mappings
P.W. Michor

University of Strathclyde Seminars in Applied Mathematical Analysis: Vibration Theory
G.F. Roach (Editor)

Notes on Real and Complex C*-Algebras
K.R. Goodearl

Applications of Combinatorics
R.J. Wilson (Editor)

University of Strathclyde Seminars in Applied Mathematical Analysis: Mathematics in Medicine and Biomechanics
G.F. Roach (Editor)

University of Strathclyde Seminars in Applied Mathematical Analysis: Multiparameter Problems
G.F. Roach (Editor)

University of Strathclyde Seminars in Applied Mathematical Analysis: Classical Scattering
G.F. Roach (Editor)

Applied mathematical analysis:

mathematics in medicine and biomechanics

University of Strathclyde Seminars in Applied Mathematical Analysis:

Mathematics in Medicine and Biomechanics

G F Roach (Editor)
University of Strathclyde

Shiva Publishing Limited

SHIVA PUBLISHING LIMITED
4 Church Lane, Nantwich, Cheshire CW5 5RQ, England

Available in North America from:
BIRKHAUSER BOSTON, INC.
P.O. Box 2007, Cambridge, MA 02139, USA

British Library Cataloguing in Publication Data

University of Strathclyde seminars in mathematics
in medicine and biomechanics.—(Shiva mathematics Series)
1. Medicine—Mathematics—Congresses
I. Roach, G.F. II. University of Strathclyde
610'.1'51 R853.M3

ISBN 0–906812–41–0
ISBN 0–906812–40–2 Pbk

Printed in Great Britain by Devon Print Group, Exeter

Contents

Preface

As a result of informal discussions during the 1975 Methoden und Verfahren der mathematischen Physik meeting at Oberwolfach it was apparent that there was an outstanding need for an international seminar group which concerned itself with problems which had been attempted but which still stubbornly resisted solutions rather than with problems which had been solved.

The first such seminar was held at the University of Bonn during the period 20-23 September, 1976 under the heading of Mixed Boundary Value Problems. Since then a number of similar meetings have been held at either the University of Bonn or the University of Strathclyde and there is now every indication that these meetings will be held regularly and, moreover, that they will become increasingly multidisciplinary. Consequently, it has been decided to make the work of these meetings more readily available by publishing proceedings. Each volume will consist of the expository lectures which it has sometimes been found necessary to include in the programme together with accounts of the research topics considered. We are most grateful to Shiva Publishing both for their agreeing to publish the proceedings as well as for their understanding and help in attempting to produce each volume as quickly as possible after each meeting.

These various meetings could not take place without a considerable amount of support and assistance from a number of sources. In this connection we are particularly grateful to

The British Council
The U.S. Army

Sonderforschungsbereich 72
The University of Strathclyde

for all the help they have given and continue to give to these seminars.

University of Strathclyde G. F. Roach
Glasgow G1 1XH

September 1983

The viscosity of blood

J.C. BARBENEL, G.D.O. LOWE and C.D. FORBES

INTRODUCTION

The blood transports nutrients and oxygen to the cells which constitute the tissues of the body, and remove metabolic waste products. It is distributed through the large vessels of the circulatory system, driven by the pressure generated by the heart. A knowledge of the gross rheological behaviour of the blood is important to the understanding of the mechanics of the circulation in health and disease.

There is also considerable interest in modelling the rheological properties of blood in terms of its constituents. Blood is not a continuum, but consists of cells suspended in a continuous fluid medium. Many of the gross rheological properties can be described, and understood, in terms of the behaviour of the subphases. These may be abnormal in disease, and this has given rise to an interest in the direct clinical relevance of blood rheology to diagnosis and to the causation of disease (see Forbes, Lowe and Barbenel, this volume).

THE COMPOSITION OF BLOOD

Blood consists of a continuous aqueous phase, the plasma, in which cells are suspended. The commonest cells are the red blood cells or erythrocytes. The volume occupied by red cells in the blood is called the haematocrit, and is usually expressed either as a percentage or a proportion. The normal haemocrit is around 45%.

The erythrocyte consists of a complex aqueous solution of haemoglobin contained within a flexible cell membrane. The membrane is highly deformable if the change in surface area is

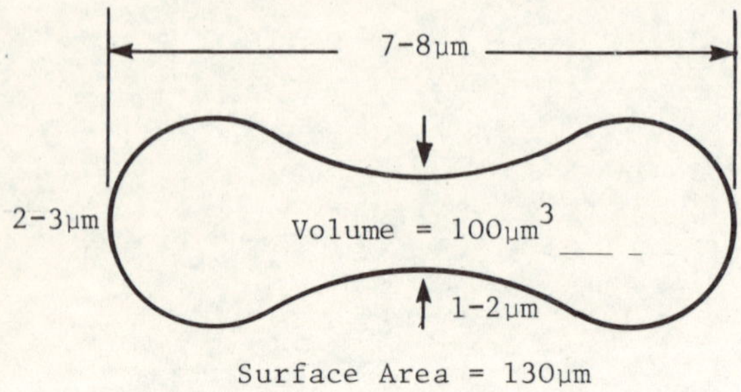

Figure 1. Cross sectional shape and typical dimensions of undeformed human red blood cell.

small, but becomes very much stiffer if the deformation produces a large change in the area of the cell membrane. The cell is not spherical; in blood at rest it is a bi-concave disc (fig. 1). The cell has a large surface area relative to its volume, which allows it to deform to produce complex shapes without producing a large areal strains. The erythrocyte in flowing blood assumes a variety of shapes and drop, cup and parachute shapes have been described and photographed (Skalak and Branemark, 1969).

The red cell surface carries a net negative charge, but in stationary blood the cells interact with each other to form aggregates. The aggregates commonly consist of 6 to 10 cells stacked face to face, and such a cluster of cells is called a rouleau. Secondary aggregation of the rouleaux also occurs building up a complex three-dimensional structure.

In normal blood cellular components other than red cells compose only about 3% of the total cell volume, (Table 1), and usually have only a small affect on blood viscosity. A suspension of leucocytes shows a higher viscosity than the same volume fraction of red cells, (Steinberg and Charm, 1971) and white cells may significantly increase blood viscosity in diseases where there is a large increase in leucocyte concentration (Dintenfass, 1965; Lichtman, 1973).

	CELLS/µl	SHAPE & SIZE	NUCLEUS
RED CELLS	5×10^6	See Figure 1	None
WHITE CELLS	7×10^3	–	–
Granulocytes	4.5×10^3	Spherical 7-9 µm	Lobed
Lymphoyles	2×10^3	Spherical 6 µm	Spherical
Monocytes	0.5×10^3		
PLATELETS	300×10^3	Rounded or oval 2-4 µm	None

Table 1. Representative values for cellular elements in human blood

Plasma is an aqueous solution of ions, and inorganic and organic molecules. The inorganic and small organic molecules contribute little to plasma viscosity which is primarily dependent on the plasma proteins. If blood is allowed to clot and the solid material removed, the remaining fluid is called serum. This has essentially the same composition as plasma except that fibrinogen and some of the clotting factors have been removed.

THE MEASUREMENT OF VISCOSITY

The coefficient of this viscosity, commonly called the viscosity (μ) of the fluid is the ratio of the shear stress (τ), and the shear strain rate ($\dot{\gamma}$) occurring at the same point in the flowing fluid (Fung, 1981). There are a large variety of methods and instruments for the measurement of viscosity (van Wazer et al, 1963) and the choice of technique depends on the general rheological properties of the fluid and special features associated with blood.

The most important general consideration is whether the fluid is Newtonian, that is, the coefficient of viscosity is independent of the shear rate. Blood is non-Newtonian, and ideally a test apparatus which subjects the sample to a uniform shear rate throughout the test volume should be used. Cone and plate viscometers have been employed (see Dormandy, 1981), but concentric cylinder geometries offer major advantages (Cokelet, 1972).

In the most common type of concentric cylinder viscometer a cylinder (radius R_b) is suspended co-axially, in a fluid sample contained in a cylindrical cup (radius R_c). The cup or the suspended cylinder - usually called the bob - is rotated at a known angular velocity Ω and the torque (M) transmitted to the bob or cup is measured. The shear stress at any radius R is given by:

$$\tau(R) = M/2\pi R^2 L$$

Where L is the length of the bob.

The shear rate for a Newtonian fluid is:

$$\dot{\gamma}(R) = \frac{2\Omega\ R_c^2 R_b^2}{(R_c^2-R_b^2)R^2}$$

Thus, the shear rate is only approximately uniform and varies across the angular gap. The variation will be reduced if the gap width is kept small. For a non-Newtonian fluid the shear rate variation is more complex and has been analysed by Krieger and Elrod (1953). The application of their work to blood is discussed by Chien (1975) and Cokelet (1972). The viscosity of the non-Newtonian fluid is commonly expressed as the apparent viscosity (μ_a), the ratio of the shear stress and mean shear rate, both measured at the same rotational speed.

$$\mu_a = \frac{\tau}{\dot{\gamma}} = \frac{(R_c^2-R_B^2)M}{4\pi R_c^2 T_B^2 L\Omega}$$

The shear rate in the fluid between the base of the bob and cup can be made to approximately match that of the fluid in the annular gap by making the ends of both the cap and bob conical. The effect of this region can be taken into account by using an effective bob length in place of the actual length.

Plasma from normal patients is Newtonian or at worst exhibits only very slight non-Newtonian behaviour. (Chien 1975). Capillary viscometers are commonly used for plasma viscosity measurements (Harkness, 1981).

Problems in Blood Viscosity Measurement

Blood consists of a suspension of particles and at the viscometer-blood interfaces of the cup and bob there is an exclusion affect. Thus, the fluid near the viscometer walls is different in cell concentration from that in the bulk sample. Cokelet et al (1963 and 1972) have suggested that the affect may not be of great importance when the hematocrit is below about 40%. The density of the red cells is greater than that of the plasma and they will settle under the influence of gravity. The speed of sedimentation is increased by the formation of the rouleaux. Sedimentation seem not to be of major importance if measurements are made in under about fifteen minutes. The erythrocytes are deformable and will migrate in flow fields in which shear stress gradients are present. Such migration may occur in concentric cylinder viscometer, and may produce major artifacts at very low shear rates, (see below).

Most investigations of blood have used anti coagulants to prevent clotting, thus extending the time available for measurement. The affect of anti-coagulants on blood viscosity appears to be small (Cokelet, 1963; Galuzzi et al 1964; Charm and Kurland 1974). The plasma proteins in blood may form an interfacial film if a blood-air interface is present. This affect may be reduced by using a cap on the viscometer cup.

The viscosity of blood and plasma like other liquids, falls as the temperature is increased. Measurements should be made at constant temperature, and 37°C, the deep body temperature, is commonly used.

GROSS RHEOLOGICAL BEHAVIOUR OF BLOOD

Normal plasma has a viscosity of 1.15 - 1.35mPas at 37°C (Harkness, 1981). The increased viscosity relative to water is produced mainly by the presence of plasma proteins. Both the concentration and the molecular shape of the proteins are important. Fibrinogen which has an elongated molecule has a marked influence; although forming less than 5% of the total plasma protein it is responsible

	CONC. g/l	MOLECULAR WEIGHT	MOLECULAR SIZE, nm
PROTEIN	75	–	–
Albumin	45	69000	15 × 4
Globulin	27	150000	24 × 4.5
Fibrinogen	3	340000	70 × 4

Table 2. Protein concentration, human plasma.

for about 20% of the plasma viscosity elevation.

Blood displays shear thinning non-Newtonian behaviour, with the apparent viscosity falling with increasing shear rates. At high shear rates the blood shows Newtonian characteristics, with a limiting viscosity of about 5mPas in normal subjects. The shear rate at which the transition to Newtonian behaviour occurs is ill defined. Merrill and Pelletier (1967) suggested a shear rate of about $100s^{-1}$, Chien et al (1966) thought that Newtonian behaviour occured about 200 to $400s^{-1}$.

Many investigators (Scott-Blair, 1959; Charm and Kurland, 1962; Merrill and Pelletier, 1967) have suggested that the variation of shear stress with shear rate can be described by the equation first derived by Casson (1959),

$$\tau^{\frac{1}{2}} = a\dot{\gamma}^{\frac{1}{2}} + \tau_y$$

The plot of the square root of the shear stress against the square root of the shear strain often produces straight lines over a considerable range of shear rates. The constant, τ_y, is equivalent to a yield stress. Unfortunately the data at low shear rates ($< 1s^{-1}$) is often unreliable, because of these shear rates blood shows time dependence, and torque relaxation occurs when the blood is subjected to constant shear rate.

The rheological behaviour of blood has also been modelled in terms of simple power law expressions of the form,

$$\tau = b\dot{\gamma}^k$$

The constant, k, generally has a value in the range 0.5 and 0.8. The expression is of restricted usefulness and although power law relationships model the behaviour over a narrow range of shear rates Whitmore (1968) has demonstrated their limitation.

The constitutive equation suggested by Herschel and Bulkley (1926),

$$\tau = b\dot{\gamma}^k + \tau_y$$

has also been used to describe the shear rate dependence of blood, Charm & Kurland (1974) reported that it, too, is appropriate for only a limited shear rate range.

The constants associated with many of the constitutive equations are often, in fact, parameters, their numerical values being a function of haematocrit.

RELATIONSHIP BETWEEN CONSTITUENTS AND VISCOSITY OF BLOOD

Haematocrit and Relative Viscosity

The blood consists of a suspension of cells in plasma, and tests show that increasing the haematocrit elevates the blood viscosity. Attempts to produce a numerical relationship between the two variables have had only limited success.

Einstein (1906) proposed that for a suspension of rigid particles the relative viscosity (μ_r) - the ratio of the viscosity of the suspension to that of the suspending medium, - was related to the volume fraction (ϕ) of particles by:

$$\mu_r = 1 + c\phi$$

c being a constant, and having a value of 2.5 for rigid spheres. The derivation assumes that the volume concentration of suspended particles is small (0.01 - 0.05), that they are rigid and that they

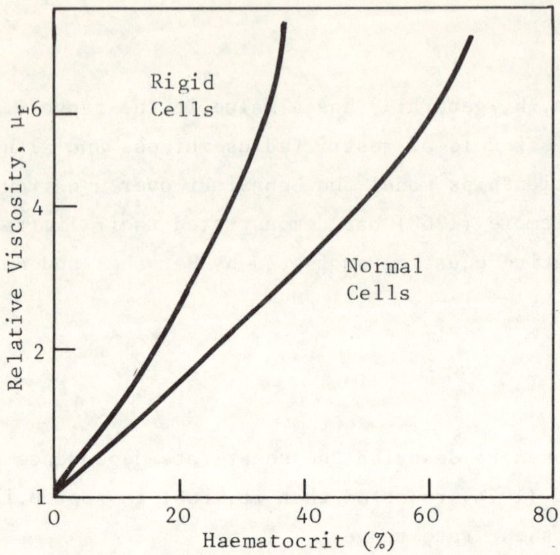

Figure 2. Influence of haematocrit on the relative viscosity of suspensions of normal and hardened red blood cells in plasma.

do not interact. None of these assumptions are valid for blood.

Empirical modifications of the Einstein equation have been suggested by numerous workers. Vand (1948) suggested that

$$\mu_r = 1 + 2.5\phi + 2.35\phi^2$$

and a similar relationship for blood has been suggested by Brooks et al (1970) relating relative viscosity to haematocrit (H),

$$\mu_r = a_0 + a_1H^2 + a_2H^3 + \ldots .$$

It was suggested that the constants $a_1, a_2, \ldots$ were related to physical mechanism occuring during flow, such as particle - particle interactions, and particle-medium interactions, but no numerical values were suggested for the coefficients.

The Einstein analysis was extended to a dilute suspension of deformable spheres by Taylor (1932). This, together with an incremental theory for the viscosity of suspensions (Brinkman, 1952; Roscoe, 1952) has formed the basis of an alternative method for predicting blood viscosity. The suspension produced by the addition of red cells to plasma is considered as a continuum with a raised viscosity. The addition of a further small amount of red cells can, therefore, be considered as the production of a new, dilute, suspension. The analysis (Dintenfass, 1968) yields a relationship of the form:

$$\mu_r = (1-kH)^{-2.5}$$

A similar theoretical analysis of Quemada (1983) led the alternative relationship;

$$\mu_r = \exp(2.5H)$$

Exponential constitutive equations have also been proposed on purely empirical grounds, to describe the linear relationship between the logarithm of viscosity and haematocrit (Gregersen et al, 1965; Begg and Hearns, 1966; Sirs, 1982).

The viscosity depends also on shear rate, as is apparent in the previous section. An equation which incorporates both shear rate and haematocrit was suggested by Wallburn and Schneck (1976).

$$\tau = a_1 \exp(a_2 H - \frac{a_3}{H^2})\dot{\gamma}^{1-a_4 H}$$

Very few of the relationships which have been suggested have been critically evaluated and the range of applicibility of the various equations is far from clear. The multiplicity of proposed constitutive equations may be due to the fact that characteristics of the blood, other than haematocrit and shear rate dependence have been omitted.

Red Cell Deformation

The deformation of red blood cells allows the blood to remain fluid at haematocrits of 98%; rigid cells will cease to flow at a cell concentration of about 60%. The fall in viscosity with increasing

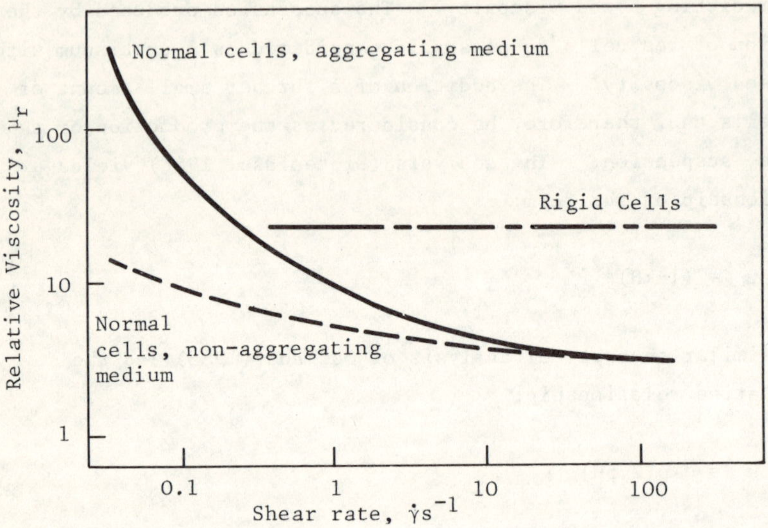

Figure 3. Variation of blood viscosity with shear rate, demonstrating the influence of red cell flexibility and aggregation. After Chien (1975).

shear rate is also a manifestation of the deformation of the erythrocytes. The effect of cell deformation can be clearly demonstrated by comparing the viscometric behaviour of normal and rigid red cells (figure 3). The rigid cell suspensions are considerably more viscous and the shear thinning behaviour is abolished.

Schmid-Schönbein et al (1973) directly observed the flow of red cells in a transparent, counter-rotating viscometer. The cells were suspended in dextran with a viscosity of about 60mPas. At a shear rate of 20s^{-1} the cells were deformed into ellipsoids, and the elongated cells were aligned in the direction of flow. Schmid-Schönbein (1976) suggested that the cell membrane rotated about the cellular contents, transmitting the shear stresses to the fluid inside the cell, which also showed rotation. It is not clear, however, that this cell content circulation occurs if the suspending medium has a lower physiological, viscosity (Chien 1975).

The haematocrit also influences the deformation of the erythrocytes. Goldsmith (1971) studied the flow of red cells and red cell ghosts (cells not containing haemoglobin) in glass capillaries. Raising the cell concentration produced an increase in both cell deformation and the alignment of the cells in the direction of the flow.

Red Cell Interaction

In stationary blood rouleaux form which interact to produce larger aggregates. At low flow rates the presence of these red cell structures strongly influences the viscosity of the blood. The size of the rouleaux and aggregates progressively decrease as the shear rate increases, and this produces the typical shear thinning seen in blood at low shear rates. In normal blood the disaggregation is probably complete at shear rates about $50s^{-1}$ (Schmid-Schönbein et al, 1968; Chien, 1975).

Rouleaux and aggregate formation is a complex and little understood process, but depends on the adsorption of macromolecules onto the surface of the red cell membrane. The adsorbed molecules form bridges between adjacent cells (Chien and Jan, 1973). The stability of the aggregates depend on the bridging forces produced by the macromolecules, the mutual repulsion of the negatively charges red cell membranes and the mechanical forces produced during flow (Chien 1975). The flexibility of the red cell is an important factor in producing stable rouleaux, and ultrastructural studies have shown deformation of adjacent red cells, which reduces the inter cell separation.

Red cells aggregate strongly in plasma, fibrinogen being a major aggregating protein. Rouleaux formation and aggregation can be abolished by suspending normal red blood cells in a fibrinogen free medium, such as serum or saline (Cokelet, 1972; Goldstone et al, 1970). This allows the experimental evaluation of the influence of aggregation factors on low shear viscosity. Comparing normal red cells in plasma and in albumin - Ringer solution at physiological haematocrits (Figure 3) shows that at shear rates below $10s^{-1}$ the relative viscosity of red cell suspensions is primarily controlled by the degree of cell aggregation (Chien 1975). The influence of cell aggregates depends on the red cell concentration. At very low haematocrits there are too few cells present to interact; at very high cell concentrations (≃ 90%) similar rheological behaviour is seen whether or not aggregating macromolecules are present, because the cells are in close proximity (Chien et al 1966). Aggregation appears to produce the greatest affect at haematocrits near or slightly higher than normal values (Usami et al, 1969).

Yield Stress

The viscosity of blood increases as the shear rate falls, but it is uncertain what happens when the shear rate actually falls to zero. The build-up of a three dimensional structure of interacting aggregates suggests that blood may show a yield stress, but this is a matter of disagreement between different workers, often using different experimental techniques and equipment.

Benis and La Coste (1968) investigated the yield stress of blood using a balance plate apparatus, which was a parallel plate viscometer measuring the stress relaxation when one plate was displaced. Yield stress values of 0.1-2mPa were found for blood with haematocrits in the range 40 - 70%. Merrill et al (1965) showed that a pressure difference could be maintained across a capillary viscometer without detectable fluid flow. Calculated wall stresses suggested yield values for blood 2-5 times greater than Benis or La Coste.

The concentric cylinder viscometers used to measure blood viscosity have also been used to investigate yield stress. Torque decay after the rotating bob has been suddenly stopped should give information from which the yield stress can be estimated. Unfortunately many of the viscometers have soft torque measuring systems with long time constants, which produce artifactual apparent torque relaxation curves. Complex analysis is required to obtain values of yield stress from such torque relaxation data (Cokelet, 1972). Estimates of yield stress may also be obtained by extrapolating torque-shear rates to zero rate. Extrapolation of Casson plots produce estimates of yield stress values which agree with those obtained from torque decay (Cokelet et al, 1963; Merrill et al, 1966).

Merrill et al (1963) also reported that the yield stress was a function of haematocrit. For individual samples of normal blood with haematocrits of up to 50%, it was found that:

$$\tau_y = A(H-H_o)^3$$

where A was a constant for each sample, and H_o a haematocrit value, ranging from 0-10%, for which the yield stress was zero.

There are major problems associated with the estimation of yield stress from extrapolated Casson plots. The shear rate, and shear stress, within the annular gap of a concentric cylinder viscometer is not uniform, but is greatest at the bob surface and least at the cup. If the yield stress of blood falls within the range of stresses occuring within the viscometer gap, it is possible to produce co-existing solid and liquid phases in the annulus (Bloor, 1982). This will make the interpretation of experimental results uncertain, and it will be unclear how well the Casson equation really characterises blood at these very low shear rates. Doubts of this kind have led Chniel (1979) to suggest that blood does not possess the yield stress, and that the extrapolation of the Casson plot to zero shear rate is unjustified.

At low shear rates the torque relaxation which is seen on stopping flow also occurs when rapid changes of strain rate are made. This time dependence of the torque output makes it extremely difficult to provide an unequivocal value of viscosity at low shear rates. It becomes important how such torque relaxation is treated; it is seldom specified in quoted low shear viscosity values.

TIME DEPENDENCE AT LOW SHEAR RATES

There are two characteristic features associated with time dependence at shear rates below $1s^{-1}$. The application of a rapid change in shear rate produces a steadily increasing torque which reaches a maximum value and then shows torque decay. The time required to reach the maximum increases with decreasing shear rate.

The slow initial rise has been modelled using linear viscoelasticity theory. MacMillan and Utterback (1979) showed that it could be qualitatively described by treating the blood as a Maxwell fluid having constitutive relations.

More complicated linear viscoelastic models for blood have also been suggested in an attempt to improve the quantitative agreement. Deutsch and Phillips (1977) investigated a four constant Oldroyd fluid and Stoltz et al (1978) analysed the response of a three parameter viscoelastic fluid. Both groups found that the constitutive constants derived from experimental results were, in fact, functions of the shear rate and that these model materials were not adequate constitutive representations for blood.

Torque relaxation at low shear rates only occurs when red cells are suspended in a medium containing aggregate macromolecules. It is abolished if the cells are rigid, or the continuous phase is plasma or saline. This suggests that the occurrence of time dependence is related to the formation of rouleaux, and Cokelet (1972) has suggested that it is produced by the appearance of a plasma film due to the exclusion of these rouleaux at the viscometer walls, followed by the migration of red cells within the viscometer gap.

The formation of plasma layers has been extensively investigated in capillary flow, and there is general agreement that a cell depleted layer exist near the tube walls. Charm and Kurland (1974) pointed out that, although blood flows with a non-uniform cell distribution across the tube, cells frequently bounce off the tube wall and they suggested that cell depleted plasma layers exist only in a statistical sense. Concentric cylinder viscometers with grooved surfaces have been used to investigate the rheological influence of the plasma layers and of cell migration. Cokelet (1972) stated that red cell migration was unaffected, but that the Vand exclusion effect was eliminated by using grooved viscometer surfaces. Higher peak torques were recorded with grooved viscometers.

Huang (1972) developed a thermodynamic based mathematical description for time dependent non-Newtonian fluids. The theory, which contains a molecular arrangement parameter, was adapted to describe the experimental torque relaxation and hystesis found with blood at low shear rates. A five parameter model was evaluated, and shown to characterise a very limited range of data. The generality and usefulness of the theory has yet to be established.

In the absence of any well established theoretical basis for treating time dependence, two empirical techniques have been used. Merrill, Cokelet and co-workers, suggest that true blood viscosity can be obtained by extrapolating the torque decay cure to zero time (or zero torque) before the application of the change in strain rate (see Cokelet, 1972). The shape of the torque decay curve at these short times will probably have been strongly affected by the transient response of the torque measuring system. Chien has acknowledged the existence of time dependence, but suggests that the decay is not a simple function of time, and has used the torque maximum for calculating the blood viscosity (see Chien et al 1966). Using the first treatment for time dependence will produce higher values for blood viscosity; this may be one of the reasons for disagreement on the utility of the Casson's plot at low shear rates, and on the validity of yield stress estimates obtained by extrapolating such graphs to zero strain rate.

CONCLUSION

Blood shows time dependent non-Newtonian behaviour, the importance of which depends on the shear rate. The flow properties of blood may be divided into three regions. At high shear rates the time independent Newtonian properties of blood have been well established, and there is a qualitative understanding of the underlying influence of red cell concentration and deformation. At low shear rates time dependence and non-Newtonian behaviour have been demonstrated, as has the influence of red cell aggregation. There is, however, no generally accepted way of treating time dependence. At very low and zero shear rates there is some disagreement about the existence of a blood yield stress. Rheological data at very low rates of shear will be strongly influenced by yield phenomena, and there is still a lack of agreement about the validity and interpretation of experimental rheological results in this region.

It is clear that there remains very considerable experimental and theoretical work to be done on the viscosity of blood.

REFERENCES

Begg, T.B., and Hearns, J.B., (1966), Components in Blood Viscosity, Clin. Sci. 31, 87-93.

Benis, A.M., and La Coste, J., (1968), Study of Erythrocyte Aggregation by Blood Viscometry at Low Shear Rates Using a Balance Method, Circ. Res., 22, 29-41.

Bloor, M.I.G., (1982), An Analysis of Blood Flow at Low Shear Rates in a Concentric Cylinder Viscometer, Biorheology, 9, 681-694.

Brinkman, H.C., (1952), The Viscosity of Concentrated Suspensions and Solutions, J. Chem. Phys., 20, 571-578.

Brooks, D.E., Goodwin, J.W. and Seaman, G.V.F., (1970), Interactions Among Erythrocytes Under Shear, J. Appl. Physiol., 28, 172-177.

Casson, N., (1959), A flow Equation for Pigment-Oil Suspensions of the Printing Ink Type, in "Rheology of Disperse Systems", Ed. Mill, C.C., Pergamon Press, Oxford.

Charm, S.E., and Kurland, G.S., (1965), Static Method for Determining Blood Yield Stress, Nature, 216, 1121-1123.

Charm, S.E., and Kurland, G.S., (1974), "Blood Flow and Microcirculation" Wiley Interscience, New York.

Chien, S., (1975), Biophysical Behaviour of Suspensions, in "The Red Blood Cell, Vol. II", Ed., D.M. Surgenor, Academic Press, New York.

Chien, S., and Jan, K.M., (1973), Ultrastructural Basis of the Mechanism of Rouleaux Formation, Microvasc.Res., 5, 155-166.

Chien, S., Usami, S., Taylor, H.M., Lundberg, J.L. and Gregersen, (1966), Effects of Haematocrit and Plasma Proteins on Human Blood Rheology at Low Shear Rates, J. Appl. Physiol., 21, 81-87.

Chmiel, H., (1979), Determination of Blood Rheological Parameters and Clinical Application, in "Adv. Cardiovasc, Phys. 3", Ed. Chista, D.N., Karger, Basel.

Cokelet, G.R., (1972), The Rheology of Human Blood, in "Biomechanics" Ed. Fung, Y.C., Perone, N. and Anliker, M., Prentice-Hall, Englewood Cliffs, New Jersey.

Cokelet, G.R., Merrill, E.W., Gilliland, E.R., and Shin, H., (1963), The Rheology of Human Blood Measurement Near and at Zero Shear Rate, Trans. Soc. Rheol., 7, 303-317.

Deutsch, S. and Phillips, W.M., (1977), The Use of the Taylor-Couette Stability Problem to Validate a Constitutive Equation for Blood, Biorheology, 14, 253-266.

Dintenfass, L., (1965), Viscosity of the Packed Red and White Blood Cells, Exp. mol. Pathol. 4, 597-605.

Dormandy, J.A., (1981), Measurement of Whole-Blood Viscosity, in "Clinical Aspects of Blood Viscosity and Cell Deformability", Ed. Lowe, G.D.O., Barbenel, J.C., and Forbes, C.D., Springer-Verlag, Berlin.

Einstein, A., (1906), Eine neue Bestimmung der Molekuldimensionen, Ann. Physik. Lpz. 19, 289-306.

Fung, Y.C., (1981), Biomechanics-Mechanical Properties of Living Tissues, Springer-Verlag, New York.

Galuzzi, W.J., DeLashnutt, R.E. and Connolly, J.J., (1964), Failure of Anticoagulants to Influence the Viscosity of Whole Blood, J. Lab. Clin. Med., 64, 773-777.

Goldsmith, H.L., (1971), Deformation of Human Red Cells in Tube Flow, Biorheology, 7, 235-242.

Goldstone, R., Schmid-Schönbein, H., and Wells, R., (1970), The Rheology of Red Blood Cell Aggregates, Microvasc. Res., 2, 273-281.

Gregersen, M.I., Branko, P., Chien, S., Sinclair, R.D., Chang, C., and Taylor, H., (1965), Viscosity of Blood at Low Shear Rates; Observations on its Relation to Volume Concentration and Size of the Red Cells, in "Fourth International Congress on Rheology, Part 4 - Symposium on Biorheology", Ed. Copley, A.L., Inter-Science, New York.

Harkness, J., (1981), Measurement of Plasma Viscosity, in "Clinical Aspects of Blood Viscosity and Cell Deformability", Ed. Lowe, G.D.O., Barbenel, J.C., and Forbes, C.D., Springer-Verlag, Berlin.

Herschel, W.H., and Bulkley, R., (1926), Measurement of Consistency as Applied to Rubber-Benzene Solutions, Proc. Am. Soc. Test. Mat. 26, Part II, 621-629.

Huang, C.R., (1972), A Thermodynamic Approach to Generalised Rheological Equations of State for Time Dependent and Time Independent Non-Newtonian Fluids, Chem. Eng. J. 3, 100-104.

Huang, C.R., Siskovic, N., Robertson, R.W., Fabisiak, W., Smitterberg, E.H., and Copley, A.L., (1975), Quantitative Characterisation of Thixotropy of Whole Human Blood, Biorheology, 12, 279-282.

Krieger, I.M., and Elrod, H., (1953), Direct Determination of the Flow Curves of Non-Newtonian Fluids II - Shearing Rates in the Concentric Cylinder Viscometer, J. Appl. Phys. 23, 147-149.

Lichtman, M.A., (1973), Rheology of Leucocytes, Leucocyte Suspensions, and Blood in Leukemia, Possible Relationship to Clinical Manifestations, J. Clin. Invest. 52, 350-358.

McMillan, D.E., and Utterback, N., (1979), Maxwell Fluid Behaviour of Blood, Biorheology, 17, 343-354.

Merrill, E.W., Benis, A.M., Gilliland, E.R., Sherwood, T.K., and Salzman, E.W., (1965), Pressure-Flow Relations of Human Blood in Hollow Fibres at Low Shear Rates, J. Appl. Physiol., 20, 954-967.

Merrill, E.W., Gilliland, E.R., Cokelet, G., Shin, H., Britten, A., and Well, E.R., (1963), Rheology of Human Blood, Near and at Zerc Flow. Effects of Temperature and Hematocrit level, Biophysical J, 3, 199-213.

Merrill, E.W., Gilliland, E.R., Lee, T.S., and Salzman, E.W.,(1966), Blood Rheology: Effect of Fibrinogen deduced by Addition, Circ. Res., 18, 437-446.

Merrill, E.W., and Pelletier, G.A., (1967), Viscosity of Human Blood: Transition from Newtonian to Non-Newtonian, J. Appl. Physiol., 23, 178-182.

Quemada, D., (1983), Blood Rheology and its Implication in Flow of Blood, in "Arteries and Arterial Blood Flow", Ed. Rodkiewicz, Cz., M., Springer-Verlag, Berlin (In Press).

Roscoe, R., (1952), The Viscosity of Suspensions of Rigid Spheres, Br. J. Appl. Phys., 3, 267-269.

Schmid-Schönbein, H.,(1976), Microrheology of Erythrocytes, Blood Viscosity, and the Distribution of Blood flow in the Microcirculation, Int. Rev. Physiol., 9, 1-62.

Schmid-Schönbein, H., Gaentgens, P., and Hirsch, H., (1968), On the Shear Rate Dependence of Red Cell Aggregation in vitro, J. Clin. Invest., 47, 1447-1454.

Schmid-Schönbein, H., Gosen, J.V., von Heinich, J., van Klose, H.J. and Volger, E., (1973), A Counter-Rotating 'Rheoscope Chamber' for the Study of the Microrheology of Blood Cell Aggregation by Microscopic Observation and Microphotometry, Microvasc. Res. 6, 366-376.

Scott-Blair, G.W., (1959), An Equation for the Flow of Blood, Plasma and Serum Through Glass Capillaries, Nature, 183, 613-614.

Sirs, J.A., (1981), Erythrocyte Flexibility and Whole-Blood Viscosity, in "Clinical Aspects of Blood Viscosity and Cell Deformability". Ed., Lowe, G.D.O., Barbenel, J.C., and Forbes, C.D., Springer-Verlag, Berlin.

Skalak, R., and Brånemark, P.I., (1969), Deformation of Red Blood Cells in Capillaries, Science, 164, 717-719.

Steinberg, M.H. and Charm, S.E., (1971), Effect of High Concentrations of Leucocytes on Whole Blood Viscosity, Blood 38, 299-302.

Stoltz, J.F., Gailland, S. and Guillot, M., (1978), An Approach to the Visco-Elastic Properties of Blood in Transient Flow, in "Biorheology", Ed., Huang, C.R., and Copley, A.L., A I Ch. E. Symposium No. 182, 74, 4-9.

Taylor, G.I., (1932), The Viscosity of a Fluid Containing Small Drops of Another Fluid, Proc. R. Soc. A, 138, 41-48.

Usami, S., Chien, S., and Gregersen, M.I., (1969), Viscometric Characteristics of Blood of the Elephant, Man, Dog, Sheep and Goat, Amer. J. Physiol., 217, 884-890.

van Wazer, J.R., Lyons, J.W., Kim, K.Y., and Colwell, R.E., (1963), "Viscosity and Flow Measurement", Wiley Interscience, New York.

Vand, V., (1948), Viscosities of Solutions and Suspensions, J. Phys. Coll. Chem., 52, 300-314.

Walburn, F.J. and Schneck, D.J., (1976), A Constitutive Equation for Whole Human Blood, Biorheology, 13, 201-210.

Whitmore, R.L., (1968), "Rheology of the Circulation", Pergamon Press, Oxford

Interpretation of low shear rate data in whole blood viscometry

M.I.G. BLOOR

ABSTRACT

The present analysis is concerned with the interpretation of low shear rate data in a concentric cylinder Couette viscometer, although conclusions appropriate to other geometries may be drawn from the results. In order to understand the form of the torque shear rate behaviour it is necessary to look at the detailed flow field within the viscometer gap. The effect of a yield stress is considered together with the formation of at least one and possibly two plasma layers adjacent to the solid boundaries. Although the Vand effect is not considered in detail it will be seen from the analysis that the inclusion of this effect would only lead to minor modifications in the theory. It will be shown that, depending on the size of a certain parameter the nature of the flow within the viscometer changes in such a way that interpretation of the data is not at all a simple matter. More tentatively, the idea of a critical viscosity and associated critical shear rate is put forward.

INTRODUCTION

In recent years there has been a great deal of effort devoted to the measurement and modelling of the non-Newtonian behaviour of human blood. The deviation from Newtonian behaviour is especially pronounced in low shear rates and is usually attributed to the aggregation of red blood cells. The cells form highly deformable rouleaux which in turn create coherent structures in the blood at rest and so give rise to a sort of yield stress. This bears some resemblance to the yield stress associated with plastic materials providing the time scale of the experiment is not too large. This point is discussed at some length by Cokelet (1) and also Fung (2) who each give an extensive review of the earlier work in this area. It is reasonable to suppose that the effect of a yield stress should be considered in any analysis of the flow in a viscometer at low shear rates. Indeed, it will be shown that this effect gives rise to experimental results which might easily be interpreted in terms of Maxwell fluid behaviour.

It is generally accepted that aggregation plays a significant role in the formation of cell free layers adjacent to solid boundaries in shear flows. It is well known that at low Reynolds number the linearity of the equations of motion imply that rigid particles are not subjected to any net transverse force in a shear flow. However, the very high deformability of the rouleaux provide the nonlinear link which gives rise to such a transverse force resulting in the depletion of red cells in the region near the wall. While it is obvious that there is no clearly defined interface between the whole blood and the plasma layer, nevertheless a dynamic equilibrium can exist. The result of this is that the whole blood is able to slip past the solid boundary on what is effectively a thin lubricating layer of plasma having a much lower viscosity than the blood. Clearly this effect must be accounted for in any accurate measurements of whole blood viscosity. Unfortunately the migration of deformable particles at the sort of concentration levels found in blood is not fully understood. However, some pointers do exist to indicate the significant parameters in the process and on this basis an experimental correction technique has been suggested by the author (3). For the present purposes it will be sufficient to take account of red cell migration on an ad hoc basis and include in the model, where appropriate, a distinct plasma layer, the thickness of which is time dependent.

A common feature of many of the attempts at developing a constitutive equation for blood has been the way in which shear stress readings have been related to overall or modified shear rate on the assumption that the material in the viscometer is homogeneous. Different types of experiment are used to highlight different aspects of the non-Newtonian behaviour. In this way characteristic times for the development of a Maxwell fluid model can be found and shown to fit the experimental data reasonably well (4, 5).

In this paper, it will be shown that the detailed flow field within the viscometer is vitally important in arriving at a correct interpretation of experimental data. Although viscoelastic effects are undoubtedly present in blood flow, such effects will not be included in the present analysis. The reason for this apparent omission is to demonstrate the appearance in experimental data of effects which appear to be rate dependent yet can be explained without reference to a viscoelastic model. Recognizing that these features can arise in more ways than one should lead to a clearer understanding of the behaviour of blood at low shear rates and also assist in evaluating the viscoelastic properties.

FLOW MODEL

The present analysis is based on the model for blood flow in a concentric cylinder Couette viscometer previously developed by the author (3). A cylindrical polar coordinate system (r, θ, z) is set up with z along the axis of symmetry. The inner surface on which the torque is measured is at $r = a$ and the cup which can be rotated is at $r = b$. For time $t < 0$ the system is at rest. At $t = 0$ the outer cylinder begins to rotate with angular velocity $\Omega(t)$ and the torque readings at the inner cylinder are taken; the rotational speed is such that shear rates below about 1-4 sec^{-1} occur in the blood. Arguments outlined in an earlier paper (3) with special reference to the case of $\Omega(t)$ = a constant indicated that the torque in the early stages was strain dependent. Furthermore a reduction in torque arises due to the development of plasma layers adjacent to the solid surfaces because the viscosity of the plasma is very much less than the viscosity of the whole blood at these shear rates.

Attention is now concentrated on the early development of the motion and for simplicity the effect of the plasma layer is ignored for the moment. The blood in the viscometer gap, in this early stage, is regarded as an elastic solid which for simplicity is assumed to obey a linear stress strain relationship in spite of the fact that large strains may arise. Such complications can easily be accommodated once the essential features of the model have been established. Also, the time required to establish quasi equilibrium conditions across the gap is very small compared with a time scale for the experiment. Under these circumstances the strain varies like r^{-2} and is therefore a maximum on the surface of the inner cylinder. Consequently as the cup continues to rotate, the solid blood will first yield at the inner cylinder to form a liquid, at time $t = t_s$ say. Thus, for some t after t_s the flow in the viscometer gap will be as shown in Figure 1. The interface between the liquid and solid phases is at $r = h(t)$ which must be determined so that the corresponding torque can be evaluated.

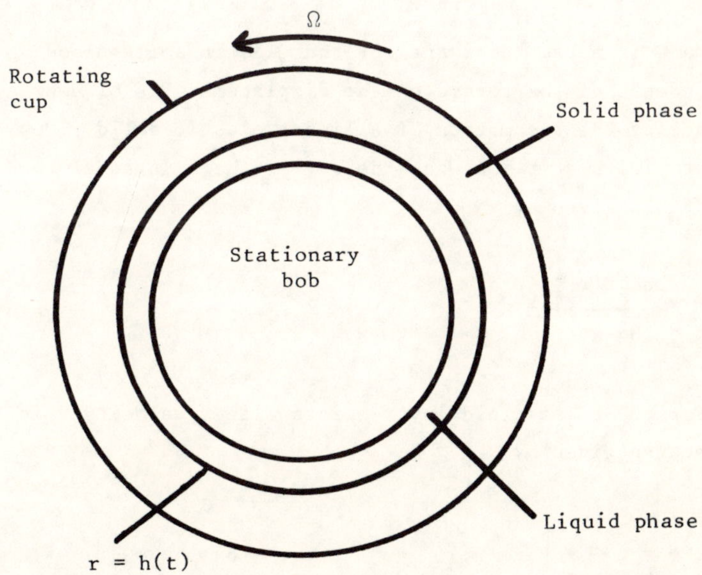

Figure 1. Diagrammatic representation of the model of the flow field.

In the solid region, denoting the azimuthal displacement of a particle by d

$$d = \frac{A_e}{r} + B_e r \tag{1}$$

where A_e and B_e are functions of time only. The details of this can be found in reference (6). The azimuthal displacement at $r = b$ is $D(t)$ and thus for $t < t_s$

$$d = \frac{a^2 b\ D}{b^2 - a^2} \left(\frac{r}{a^2} - \frac{1}{r}\right) \tag{2}$$

which satisfies the conditions that $d = D$ at $r = b$ and $d = 0$ at $r = a$; it is assumed that the deflection of the inner cylinder is negligible. The torque per unit length T on the inner cylinder is given by

$$T = \frac{4\pi G a^2 b\ D}{b^2 - a^2} \tag{3}$$

where G is a Lamé constant for the elastic solid blood. The linear dependence of the torque on the displacement is of course a consequence of the assumption of a linear elastic solid. Now, the solid first yields at $r = a$ when the stress there is equal to K and therefor t_s is given by

$$\frac{2Gb\ D(t_s)}{b^2 - a^2} = K \tag{4}$$

For $t > t_s$, a fluid layer exists with a velocity v in the θ direction given by

$$v = \frac{A_f}{r} + B_f r \qquad\qquad a \leqslant r < h\ . \tag{5}$$

where A_f and B_f are functions of time only and it has been assumed that the variation in the fluid viscosity across the thin layer is not significant. Hence, for the case when both solid and liquid phases coexist the functions of time in equations (1) and (5) can be evaluated from the boundary conditions. At $r = b$ the displacement $d = D$ and therefore

$$D = \frac{A_e}{b} + B_e b \ . \tag{6}$$

At $r = h$ the velocity and stress are continuous across the interface, so that

$$\frac{\dot{A}_e}{h} + \dot{B}_e h = \frac{A_f}{h} + B_f h \ , \tag{7}$$

where the dot denotes differentiation with respect to t, and

$$GA_e = \mu A_f \ . \tag{8}$$

Also at $r = h$ the stress is equal to the yield stress K, i.e.

$$-\frac{2GA_e}{h^2} = K \ . \tag{9}$$

Finally, at $r = a$ the velocity is zero

$$0 = \frac{A_f}{a} + B_f a \ . \tag{10}$$

Writing $k = K/2G$ and $\sigma = G/\mu$ equations (6), (8), (9) and (10) can be solved for A_e, B_e, A_f and B_f in terms of h. Substituting these values into equation (7) yields the following differential equation for h(t),

$$2k\dot{h}(1-h^2/b^2) = h[\Omega + \sigma k(1-h^2/a^2)] \tag{11}$$

It is clear from equation (11) that two separate cases must be considered, namely $\sigma k(b^2/a^2 - 1) \gtrless \Omega(t)$.

Firstly the case

$$\sigma k(b^2/a^2 - 1) < \Omega(t) \tag{12}$$

is considered when the whole of the annular region eventually becomes fluid and indeed $\dot{h}$ becomes infinite at $r = b$. Under these circumstances, the mismatch between the left hand and right hand sides of the equation (11) as $h \to b$ is due to the fact that the velocity distribution given by equation (5)relies on the equilibrium Couette flow being established on a negligible time scale. This is a perfectly reasonable approximation providing $\dot{h}$ is not too large. However as $h \to b$, $\dot{h}$ becomes infinite and it is necessary to consider the non equilibrium flow in the fluid. In other words, a Rayleigh layer forms at $r = b$ and the vorticity diffuses throughout the fluid layer until equilibrium Couette flow is again established. However, the time scale for this viscous diffusion is $O(\rho(b-a)^2/\mu)$ which for typical values of $b - a$ and μ is less than 10^{-2} seconds, i.e. very small compared with a time scale for the experiment. Indeed, the dynamics of the measuring system would fail to respond accurately to an effect on this time scale. As far as the theory is concerned it is reasonable to switch from a two layer model to a fully developed flow model instantaneously when $h = b$ at $t = t_f$ say. For $t_s < t < t_f$, the torque T is given by

$$T = 2\pi K h^2 , \tag{13}$$

using equation (9) and integration of equation (11) satisfying the condition that $h = a$ at $t = t_s$ gives $h(t)$.

For $t > t_f$, fully developed Couette flow exists in the gap between the cylinders and thus T is given by

$$T = \frac{4\pi\mu\Omega a^2 b^2}{b^2 - a^2} \tag{14}$$

the value of μ in this expression will almost certainly not be the same as the value the fluid had when the solid phase was present.

Secondly, the case

$$\sigma k(b^2/a^2-1) > \Omega(t) \tag{15}$$

must be considered. It can be seen from equation (11) that $\dot{h}$ will be zero when h attains the value $a\sqrt{1 + \Omega/\sigma k}$ (=s) and in this case $a < s < b$ because of the inequality (15). This means that the blood adjacent to the outer cylinder wall is simply strained and does not flow so that two phases are present. A failure to recognise this situation would clearly lead to a false interpretation of the experimental data.

The types of difficulties that might arise are best illustrated by reference to two specific cases. When $\Omega(t) = \Omega_o$, a constant, torque time curves have been calculated previously (3) and typical examples are shown in Figure 2. For both curves shown, the torque rises linearly from zero at $t = 0$ to $2\pi Ka^2$ at $t = t_s$ as a consequence of the linear elasticity. In the higher shear rate case, the torque then rises quite rapidly to $2\pi Kb^2$ at $t = t_f$ when the whole of the blood becomes liquid and the torque jumps to the value given by equation (14). Thereafter, in the absence of red cell migration, it remains steady. The lower shear rate curve continues to rise after $t = t_s$ but approaches the value $2\pi Ks^2$ exponentially slowly.

In both of these cases it can be seem that a failure to recognise the two phase nature of the flow might lead to quite erroneous conclusions about the physical properties of the material in terms of a Maxwell fluid. This point is further demonstrated in the next example.

In trying to evaluate time constants for a Maxwell fluid, it is often convenient to subject the fluid to an oscillatory strain rate. It is therefore appropriate to examine the type of response to be expected on the basis of the present model when $\Omega = \Omega_o \cos \omega t$ say. Here again, two distinct types of flow can arise depending on the inequality (15). However, the distinctive features associated with oscillatory motion can be demonstrated by considering just one case, namely $\sigma k(b^2/a^2 - 1) > \Omega_o$.

In this case, t_s is given by

$$\frac{b^2\Omega_o}{(b^2-a^2)\omega} \sin \omega t_s = k \tag{16}$$

using equation (4) and it has been assumed that ω is sufficiently small so that equation (16) has a real solution. In other words the displacement of the outer cylinder is sufficiently large to cause the whole blood to yield. For $t < t_s$, the torque is given by equation (3) and is therefore in phase with the displacement. For $t > t_s$, two phases exist and equation (11) must be integrated numerically so that the torque can be found using equation (13) At some stage when $t > t_s$, the right hand side of equation (11) and consequently $\dot{h}$ becomes zero and $h = h_m$ say.

This complication is more difficult to assess than might at first be supposed. Clearly, there is no difficulty in proceeding with the analysis on the basis already laid down, but another mechanism concerning the structure of blood comes into play of which account must be taken. The blood in the region $a < r < h_m$ has yielded and flowed at some stage. The arguments given in the Introduction to support the conjecture that a yield stress exists indicate that the blood just having flowed will almost certainly have a different yield stress from K. The time scale for the formation of the aggregates is of course the significant parameter here. If the time scale is negligibly short then there will be negligible change in the yield stress. On the other hand, a very long time scale compared with the duration of the experiment means that the blood having once yielded would remain in a liquid state. It is instructive to consider these two extreme cases in the absence of a model for the aggregation process. First of all, dealing with the case of when K is unchanged, equation (11) can be integrated until $h = a$ when all of the blood again becomes solid and equation (4) then determines the torque. A typical result for this calculation is shown in Figure 3. Also shown in Figure 3 is the other extreme case when the region $a < r < h_m$ remains liquid. In this case equations (6), (7), (8) and (10) determine a differential equation for A_e in the form

$$\dot{A}_e + \sigma A_e \frac{(h_m^2/a^2 - 1)}{(1 - h_m^2/b^2)} = \Omega_o \cos \omega t \frac{h_m^2}{1 - h_m^2/b^2} \tag{17}$$

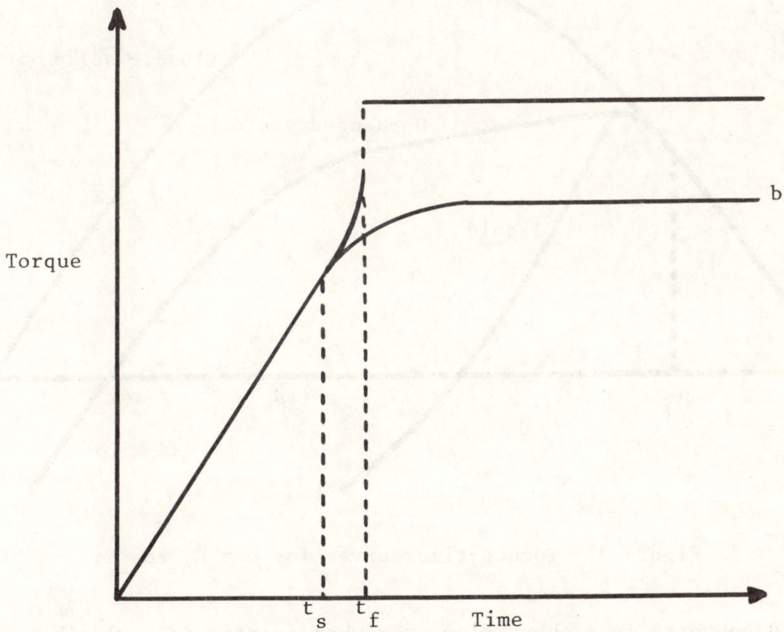

Figure 2. Typical torque-time curves for $\Omega = \Omega_o$ and (a)$\sigma k(b^2/a^2 - 1) < \Omega_o$ (b)$\sigma k(b^2/a^2 - 1) > \Omega_o$

and hence the torque is obtained. In both of these cases, the significant feature is the development of a phase difference between the torque and the displacement. It must be emphasised that this is not due to diffusion processes but to the two phase nature of the flow. A simple Maxwell fluid would of course give rise to a phase difference in this type of experiment. The difficulty in interpreting the data is again illustrated by this example.

Effect of formation of a plasma layer

It is well established that deformable particles migrate away from

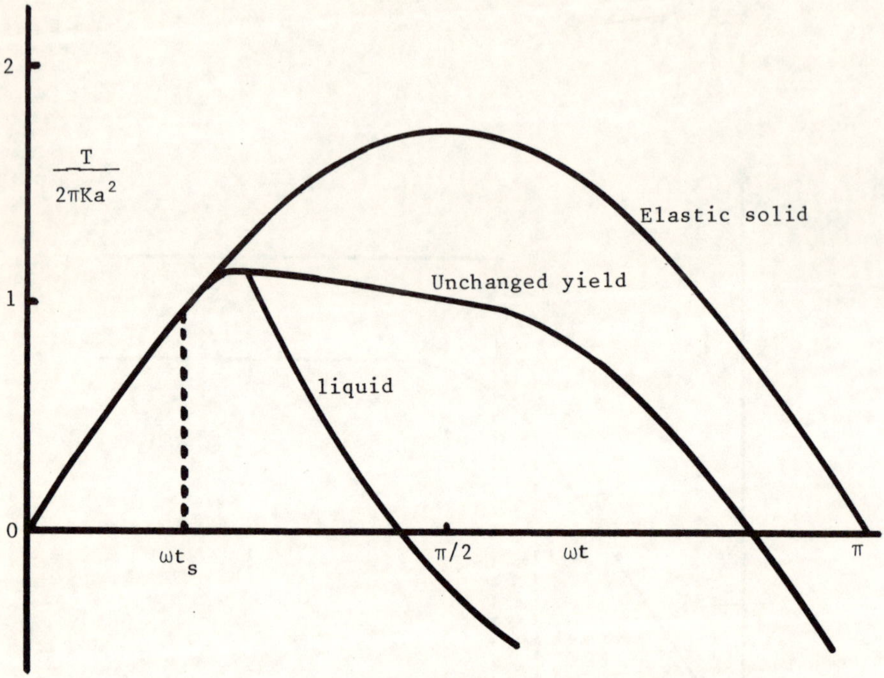

Figure 3. Torque-time curves for $\Omega = \Omega_o \cos \omega t$

a solid boundary in a shear flow even when inertia is negligible. In the case of the flow of blood in a viscometer the generation of an effectively cell free layer adjacent to a solid surface can have a significant effect on the torque required to maintain the motion. In the present problems, allowance must be made for at least one plasma layer after time $t = t_s$. Following the method established by the author (3), for $t > t_s$, but with a solid phase present in the viscometer gap, the flow can be represented in the form shown in Figure 1 with the addition of a plasma layer in the region $a < r < H$. In order to make some progress in understanding the effect of this plasma layer it is reasonable to adopt an ad hoc approach to the specification of H. If the ultimate thickness of the plasma layer is εa and it is assumed that this is approached exponentially, then

$$H = a + \varepsilon a\{1-\exp[(t_s-t)/t_o]\} \quad . \tag{18}$$

The velocity V_p of the plasma in the θ direction is given by

$$V_p = \frac{A_p}{r} + B_p r \quad . \tag{19}$$

The application of the boundary conditions (6)-(9) together with the conditions of continuity of stress and velocity at the blood/plasma interface and the no slip condition at r = a yields the following differential equation for h(t)

$$2k\dot{h}(1-h^2/b^2) = h\{\Omega + \sigma k[1-h^2/a^2(1 + 2(\nu-1) \times (H/a-1)]\}. \tag{20}$$

In deriving equation (20) use has been made of the fact that (H/a - 1) is small compared with unity and squares of this quantity have been neglected. Also, the parameter ν is the ratio μ_b/μ_p where μ_p is the viscosity of the plasma.

Comparing equation (20) with equation (11) it can be seen that the effect of the plasma layer is to reduce $\dot{h}$ by the inclusion of the extra factor $2(\nu-1)(H/a-1)$. Although (H/a-1) is small compared with unity, this extra factor will certainly be significant at low shear rates due to the large value of the viscosity ratio ν. Figure 4 shows the effect of the plasma layer on the output torque time curve for the case $\Omega = \Omega_o$. The reduction in $\dot{h}$ is evident and the qualitative effect of the presence of a plasma layer in any of the cases considered in this paper can be assessed from the result shown. Detailed calculations can be found in reference (3).

CONCLUSIONS

The model for whole blood presented in this paper is capable of explaining the peculiar features generally associated with torque time viscometer curves at low shear rates. The fact that, at sufficiently low shear rates, not all of the blood sample in the viscometer flows, makes interpretation of the data extremely

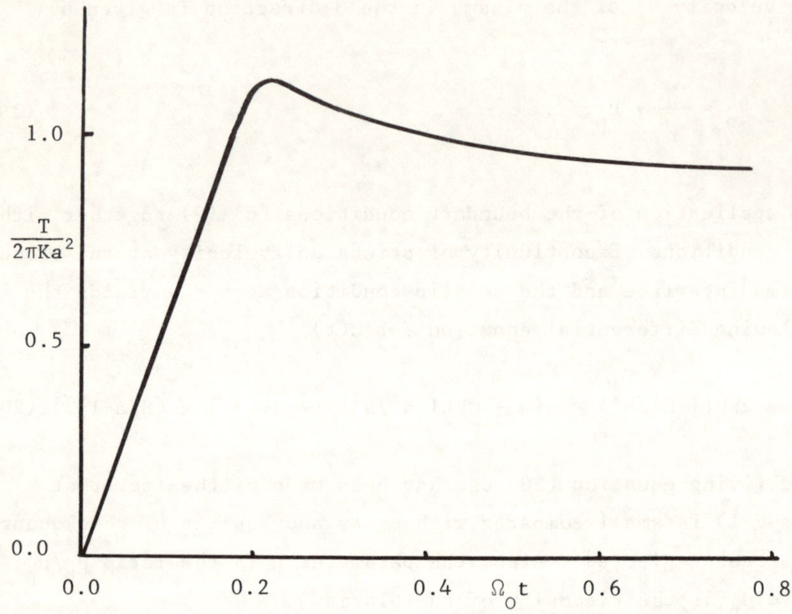

Figure 4. Torque-time curve for $\Omega = \Omega_o$, $k = 1$, $\sigma = 5$, $b/a = 1.1$ $2(\nu-1)\varepsilon = 0.3$ and $t_o = t_s$.

difficult. Another effect of the two phase nature of the flow is the appearance of what might be termed 'rate effects' in the experimental results. These arise because the torque is dependent on the relative thicknesses of the solid and liquid layers which in turn depend on the displacement and angular velocity of the cup. Interpretation of this in terms of a Maxwell fluid is not necessarily accurate and indeed tests designed to measure rate constants are open to mis-interpretation. It seems that great care should be taken in designing experimental tests to measure viscometric properties of whole blood and the fact that two phases might be present in the test should be considered when the data is analysed.

REFERENCES

1. COKELET, G.R. The rheology of human blood. In: Biomechanics Its Foundations and Objectives, Y.C. Fung (ed.) Englewood Cliffs, Prentice Hall Publ. 1972, pp. 63-103.
2. FUNG, Y.C. Biomechanics: a survey of the blood flow problems. In: Advances in Applied Mechanics, C.S.Yih (ed.) Academic Press, New York, 1971, 65-130.
3. BLOOR, M.I.G. An analysis of blood flow at low shear rates in a concentric cylinder viscometer. Biorheology, to be published.
4. THURSTON, G.B. Rheological parameters for the viscosity, viscoelasticity and thixotropy of blood. Biorheology, 16, 149-162, 1979.
5. McMILLAN, D.E. and UTTERBACK, N. Maxwell fluid behaviour of blood at low shear rate. Biorheology, 17, 343-354, 1980.
6. FILONENKO-BORODICH, M. Theory of Elasticity. Dover Publications Inc. New York, 1965, 191-194.

Malcolm I G Bloor, B.Sc Ph.D
Department of Applied Mathematical Studies
University of Leeds
LEEDS LS2 9JT

The systematic formulation of population models with distributed maturation period

S.P. BLYTHE, R.M. NISBET and W.S.C. GURNEY

1. INTRODUCTION

In this second paper on "Population Dynamics", we outline an extension to Nisbet and Gurney's[1,2] "stage structure" models of laboratory insect populations. A less mathematical treatment of the same work has been published [3], and a full account can be found elsewhere[4,5]

A principal assumption (implicit or explicit) of most age-structured population models is that ageing is uniform, so that individuals born synchronously will pass through successive developmental stages synchronously. Unfortunately, Nature is uncooperative in this respect: any given group of insect eggs, for example, will give rise to adults after a variety of total development times (see for example Kirk[6] for the duration-distribution of successive stages in the life-history of the Ground Beetle _Pterostichus chalcites_ (Say)). This is of course a particular case arising from a more general phenomenon, namely that all biological characters possess some variability, due both to the genetic differences between individuals, and to the results of differential experience during life.[7] It was to gain some understanding of the population dynamic consequences of incorporating _genetic_ variation that we considered differential ageing in laboratory insect populations.

There is to be found in the literature[8,9,10,11] an

equation purporting to be an expression of the dynamics of a sub-population of adults ($N_A(t)$) when there is some distribution ($u(a)$) of maturation periods (i.e. the time spent in all pre-adult stages):

$$\dot{N}_A(t) = F\left[\int_0^\infty u(a)N_A(t-a)da, \quad N_A(t)\right] \tag{1}$$

$F(\cdot,\cdot)$ is some non-linear function. We shall demonstrate in Section 2 that despite the popularity of (1), it is in fact quite inapplicable to age-structure modelling.

2. MODEL FORMULATION

To avoid the pitfalls of ad hoc equations such as (1), we wish to find a systematic way of formulating models of differentially ageing populations, equivalent to that[2] for uniform ageing, based on the MacKendrick-von Foerster equation[12,13]. We first note that the age of an individual, i.e. time since birth, will not tell us what developmental stage the individual is in; thus we need to define a separate "time-in-stage" for each developmental stage. For the simplest case, where there are but two developmental stages, immatures and adults, we define r and s, respectively, as the times since entering each stage. Then $f_I(r,t)dr$ and $f_A(s,t)ds$ are the numbers of immatures and adults in the respective "time-in-stage" intervals $(r,r+dr)$ and $(s,s+ds)$, respectively, at time t. The population dynamics can then be expressed by the pair of partial differential equations,

$$\frac{\partial f_I(r,t)}{\partial t} + \frac{\partial f_I(r,t)}{\partial r} = -\Phi(r)f_I(r,t) - \delta_I(r,t)f_I(r,t) \tag{2}$$

and

$$\frac{\partial f_A(s,t)}{\partial t} + \frac{\partial f_A(s,t)}{\partial s} = -\delta_A(s,t)f_A(s,t), \quad (3)$$

r, s > 0, with their associated renewal conditions, for the total birth rate,

$$B(t) \equiv f_I(0,t) = \int_0^\infty \beta(s,t)f_A(s,t)ds + I_1(t) \quad (4)$$

and for the total recruitment rate to the adult population,

$$R(t) \equiv f_A(0,t) = \int_0^\infty \Phi(r)f_I(r,t)dr + I_2(t) \quad (5)$$

$$= M(t) + I_2(t) \quad (6)$$

where $\delta_I(r,t)$ and $\delta_A(s,t)$ are the per capita death rates for the immature and adult populations, respectively, $\beta(s,t)$ is the per capita fecundity, and $\Phi(r)$ is the important function, the per capita maturation rate of immatures, i.e.

$\Phi(r)dr$ = "average fraction of the individuals in the immature stage-interval $(r, r+dr)$ who mature into adults during that interval." (7)

and $M(t)$ is thus the <u>total</u> maturation rate out of the immature stage. $I_1(t)$ and $I_2(t)$ are the initial innoculation rates to the immature and adult stages, respectively. They will both be zero over all time except for a small period just after t = 0.

We integrate equations (2) and (3) with respect to r, s over $[0,\infty)$, and obtain the integro-differential equations for total immature ($N_I(t)$) and total adult ($N_A(t)$) populations

$$\dot{N}_I(t) = B(t) - M(t) - D_I(t) \qquad (8)$$

$$\dot{N}_A(t) = R(t) - D_A(t) \qquad (9)$$

where clearly

$$N_I(t) = \int_0^\infty f_I(r,t)dr, \text{ and } N_A(t) = \int_0^\infty f_A(s,t)ds \qquad (10)$$

and where $D_I(t)$ and $D_A(t)$ are the <u>total</u> death rates for the immature and adult populations, respectively.

Equations (8) and (9) are clearly no more than a restatement of the P.D.E.s (2) and (3), to no obvious advantage, and we shall not dwell upon the completely general case. Instead, we shall concentrate upon the particular case where only the adults compete, so that there is no density-dependence in the immatures' dynamics, and no s-dependence for the adults; a competition regime that has recently received considerable attention[14,15]. Hence we can write

$$D_I(t) = \int_0^\infty \delta_I(r) f_I(r,t)dr, \qquad (11)$$

$$D_A(t) = \delta_A(N_A(t))N_A(t), \qquad (12)$$

$$B(t) = \beta(N_A(t))N_A(t) + I_1(t), \text{ and} \qquad (13)$$

$$R(t) = \int_0^\infty \Phi(r) f_I(r,t)dr + I_2(t) \qquad (14)$$

Now the solution to the P.D.E. (2) for $t > r$ is simply

$$f_I(r,t) = B(t-r)\exp\{-\int_0^r [\delta_I(x) + \Phi(x)]dx\}. \qquad (15)$$

Provided that the laboratory experiment to be modelled was started by the innoculations $I_1(t)$ and $I_2(t)$ into an empty system, then the solution to (2) for $r > t$ is zero, and we need consider only (15). By substituting equation (15) for $f_I(r,t)$ in the integral in equation (14), we remove all explicit dependence upon the immature population from the adults' population dynamic equation (9), which can thus be considered on its own, and takes the form

$$\dot{N}_A(t) = \int_0^\infty w(r)B(t-r)dr - D_A(t) + I_2(t) \qquad (16)$$

where

$$w(r) = \Phi(r)\exp\{-\int_0^r [\delta_I(x) + \Phi(x)]dr\}. \qquad (17)$$

Equation (16) is thus the correct expression of the dynamics of a sub-population of adults, where individuals pass through an immature stage whose duration is distributed according to the weighting function $w(r)$. Only under the most restricted (and unlikely) of circumstances is the "conventional" equation (1) equivalent to the desired equation (16), and there seems to be no reason for the continued use of the former. As it stands however, equation (16) does not look encouraging as a biological model; the functions $B(.)$ and $D_A(.)$ pppresent no new problems (their density-dependence has been studied, and their feasible functional forms are well known[15]), but the presence of the complicated weighting function $w(r)$ (equation (17)) makes equation (16) intimidatingly intractable. In the following Section we show that, at the cost of some biological detail, we can nonetheless make equation (16) into an analytically and computationally tractable general model.

3. THE WEIGHTING FUNCTION

Even if explicit functions for $\delta_I(r)$ and $\Phi(r)$ are available, w(r) is still sufficiently complicated to make analysis of the balance equation (16) extremely difficult. However, if we are willing to sacrifice a little detailed biological information, we can recast w(r) in more tractable form without loss of mathematical rigour. If we write

$$P(r) = \exp\{-\int_0^r [\delta_I(x) + \Phi(x)]dr\} \quad (18)$$

then

$$w(r)dr = \Phi(r)dr\ P(r) \quad (19)$$

and clearly for an individual

$$P(r) = \text{Prob (surviving, unmatured, to enter interval } (r,r+dr)). \quad (20)$$

Restating $\Phi(r)dr$ as a probability,

$$\Phi(r)dr = \text{Prob (maturing during interval } (r,r+dr) \mid \text{entering interval } (r,r+dr)) \quad (21)$$

so that

$$w(r)dr = \text{Prob (entering, and maturing during, interval } (r,r+dr)) \quad (22)$$

For a synchronously laid group (cohort) of eggs, w(r)dr is thus the fraction of the initial numbers expected to mature during the "time-in-stage" interval (r,r+dr). Defining

$$S = \int_0^\infty w(r)dr \leqslant 1 \quad (23)$$

as the fraction of the initial cohort which <u>ever</u> attains maturity (note that $S \leqslant 1$ due to the $\delta_I(r)$ term in (17)), we can write

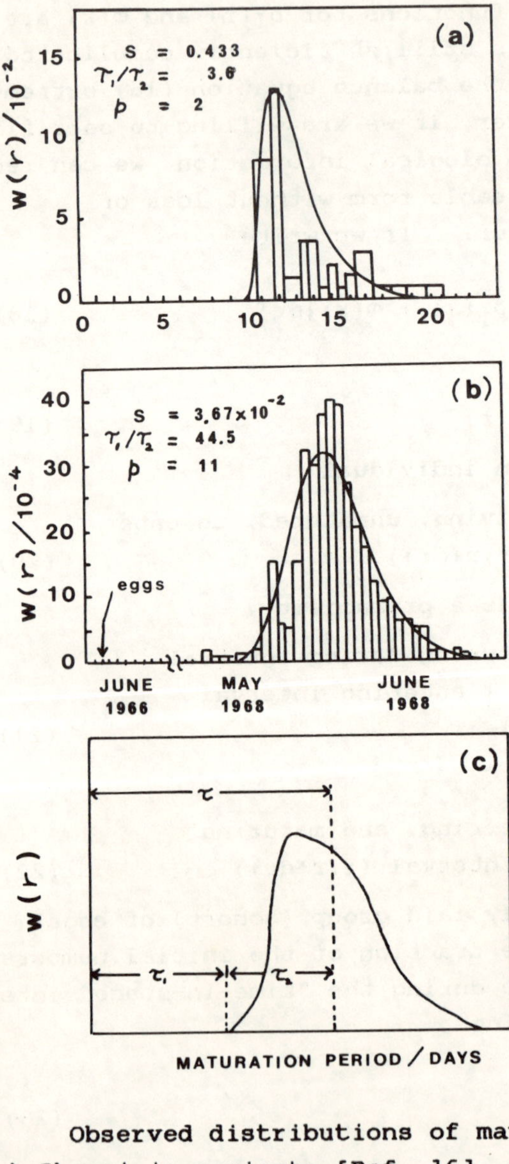

Fig. 1 Observed distributions of maturation times for: (a) Phaenicia sericata [Ref. 16], and (b) Pyrrhosoma nymphula [Ref. 17]. (c) the main features of w(r) :- the mean maturation period τ, the minimum maturation period τ_1, and $\tau_2 = \tau - \tau_1$.

$$\rho(r)dr \equiv \frac{1}{S}\, w(r)dr \tag{24}$$

$$= \text{"fraction of those ever maturing who do so during the interval } (r, r+dr)\text{."} \tag{25}$$

Hence the weighting function $w(r)$ can be replaced by a probability density function $\rho(r)$ times a scaling factor S,

$$w(r) = S\rho(r) \tag{26}$$

This compensates for the loss of detailed information on $\delta_I(r)$ and $\Phi(r)$ only if equation (16) is made tractable by the substitution of equation (26) for $w(r)$. We must thus find a form for $\rho(r)$ that is not only biologically plausible, but also facilitates the analysis of equation (16).

We observe from experimental data on maturation times that distributions are often like those illustrated in Fig. 1(a,b), for the blowfly Phaenicia sericata (Meigen)[16] and the damselfly Pyrrhosoma nymphula (Sulzer)[17], and summarised in Fig. 1(c).

There is usually some minimum duration of the immature stage (τ_1), followed by a fairly regular distribution around the mean maturation period τ. We choose a general function which displays these features, namely the gamma distribution of integer order (the Erlang distribution), with the origin shifted to $r = \tau_1$:

$$w(r) = SH(r-\tau_1)g(r-\tau_1;p) \tag{27}$$

where

$$g(x;p) = \frac{c^{p+1}}{p!}\, x^p \exp(-cx) \tag{28}$$

and $H(x)$ is the Heaviside step-function

$$H(x) = \begin{cases} 0, & x \leqslant 0 \\ 1, & x > 0 \end{cases} \tag{29}$$

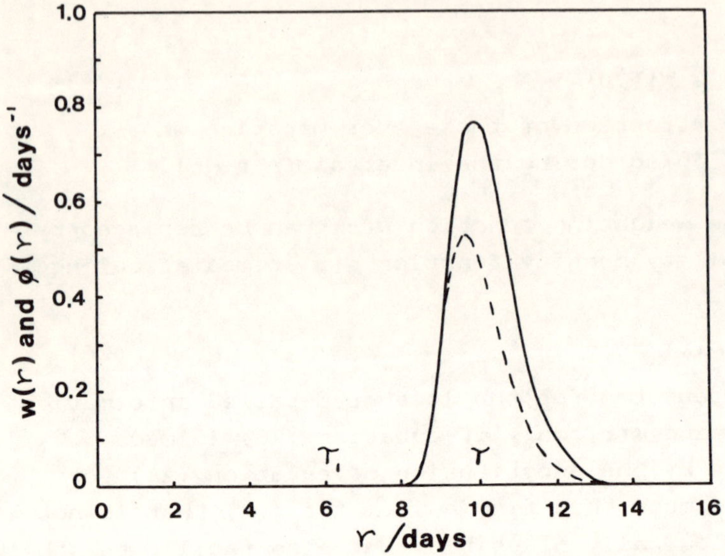

Fig. (2) A form for Φ(r) (———) which produces a w(r) (---------) in the "shifted-gamma" family (27), with p = 5, τ_1/τ_2 = 4.0, τ_1 = 8 days, $\delta_I(r)$ = 0.01 $days^{-1}$ and S = 0.5 . Obtained by solving the system of three differential equations (30) with δ(r) = 0.01, and Φ(r) = 0 for r ≤ τ_1, using the algorithm of Ref. 19.

introduced so that w(r) is described by a single equation. With suitable parameter values, equation (27) gives an acceptable characterisation of some observed w(r) (see Fig. 1 (a,b)). This "humped" form of w(r) is of course by no means universal, but seems to be sufficiently common as to be considered the norm.

There are not usually analytic expressions for Φ(r) an $\delta_I(r)$ which will result in w(r) that is exactly given by equation (27). However, by treating equation (17) as the system of three differential equations

$$\frac{dw(r)}{dr} = \Phi(r) \frac{dP(r)}{dr} + P(r) \frac{d\Phi(r)}{dr} , \qquad (30)$$

we can use a given $w(r)$ from equation (27), and some $\delta_I(r)$, to find the implied form of $\Phi(r)$.

In Fig. 2 we have plotted $w(r)$, $\delta_I(r)$ and $\Phi(r)$ for a particular case (see the caption for Fig 1 for details). Clearly, $\Phi(r)$ has a perfectly plausible form, and we need have no further qualms about the use of (27) in the analysis of equation (16).

4. MODEL ANALYSIS

Having established that the "shifted-gamma" form for $w(r)$ is acceptable, we now consider the analysis of the integro-differential equation (16). The functional forms of $B(\cdot)$ and $D_A(\cdot)$ under conditions of adult-only competition have been discussed elsewhere,[15] so that here we need only state that we expect a smoothly increasing death rate,

$$\begin{cases} D_A(0) = 0 \\ D_A(N_A) > 0 \quad \text{and} \quad D_A'(N_A) > 0 \quad \text{for} \quad N_A > 0 \end{cases} \tag{31}$$

and a birth rate that either rises smoothly to a sustained maximum

$$\begin{cases} B(0) = 0 \quad ; \quad B(N_A) > 0 \quad \text{for } N_A > 0 \\ B(N_A) \to B_{max} \quad \text{and} \quad B'(N_A) \to 0 \quad \text{as } N_A \to \infty \end{cases} \tag{32}$$

or turns over at high densities to form a "hump",

$$\begin{cases} B(0) = 0 \quad ; \quad B(N_A) > 0 \quad \text{for } N_A > 0 \, , \\ B'(N_A) > 0 \quad \text{for} \quad 0 < N_A < N_0 \\ B'(N_A) < 0 \quad \text{for} \quad N_A > N_0 \end{cases} \tag{33}$$

depending upon particular biological considerations[14,15].

Given $B(.)$ and $D_A(.)$ as described by equations (31) - (33), there is always a trivial solution to (16) at $N_A = 0$, and normally, for some region of model parameter-space, a steady-state solution N_A^*, such that _if_ $N_A(t) = N_A^*$ for $-\infty < t \leqslant 0$, _then_ $N_A(t) = N_A^*$ for $t > 0$. The linearised equation for small perturbations around N_A^* is readily accessible to a local stability analysis, and we find that the effect of increasing the mean delay τ, or of narrowing the distribution (by increasing p and/or the ratio τ_1/τ_2) is to reduce the region of parameter space where the linear equation has stable solutions. In the limit of $w(r)$ a delta-function at $r = \tau$, the distributed-delay system becomes a discrete-delay differential equation that has been much studied[15,18].

The use of the "shifted-gamma" also greatly facilitates the numerical analysis of equation (16). Details are to be found elsewhere[4,5], but essentially (p+1) auxillary variables are defined,

$$V_j(t) = \int_{\tau_1}^{\infty} g(r-\tau_1;p+1-j)\ B(N_A(t-r))\ dr\ , \qquad j = 1, 2, \ldots p+1 \tag{34}$$

and differentiated with respect to time, to give a system of (p+2) equations (similar to the "linear chain" of Macdonald[11]), which are readily solvable using an appropriate algorithm[19].

5. CONCLUSIONS

We have thus derived a single integro-differential equation describing the dynamics of a population of adults, under conditions of "adult-only" competition, when there is a distribution of immature-stage durations. This equation, fundamentally different from the conventional "distributed-delay" models, is made both mathematically and computationally more tractable by the choice of the "shifted-gamma" distribution as the weighting function.

6. REFERENCES

1. R.M. Nisbet and W.S.C. Gurney "The Systematic Formulation of Population Models for Insects with Dynamically Varying Instar Durations", Theor. Pop. Biol., - to appear
2. R.M. Nisbet and W.S.C. Gurney "Tractable Population Models with Age-Structure", - this volume
3. S.P. Blythe, R.M. Nisbet and W.S.C. Gurney "Formulating Population Models with Differential Aging", submitted to Proc. Internat. Conf. on Pop. Biol. (Edmonton, 1982), Springer-Verlag - to appear
4. S.P. Blythe "Simple Age-Structure Models of Laboratory Insect Populations", Ch.4, Ph.D. Thesis, University of Strathclyde (1982)
5. S.P. Blythe, R.M. Nisbet and W.S.C. Gurney "Population Models with Distributed Maturation Period" - in preparation
6. V.M. Kirk " Biology of Pterostichus chalcites, a Ground Beetle of Cropland", Annals Entomol. Soc. America, **68**, 855-858 (1975)
7. M. Begon "The Density of Populations and the Quality of Individuals", Presented at a Symposium on Evolutionary Ecology, Leeds 1982, - to appear
8. R.M. May "Stability and Complexity in Model Ecosystems", (2nd. Edition), Ch. 4, Princeton, University Press, Princeton, U.S.A. (1974)

9. J.M. Cushing "Integrodifferential Equations and Delay Models in Population Dynamics", Lecture Notes in Biomathematics, No. 20, Springer-Verlag, Berlin/Heidelberg/New York (1977)

10. G.F. Oster "Lectures in Population Dynamics", pp 149-190 in R.C. Di Prima (Ed.) "Modern Modelling of Continuum Phenomena", American Math. Soc., Providence, Rhode Island, U.S.A. (1976)

11. N. Macdonald "Time Lags in Biological Models", Lecture Notes in Biomathematics No. 27, Springer-Verlag, Berlin/Heidelberg/New York, (1978)

12. A.G. MacKendrick "Applications of Mathematics to Medical Problems", Proc. Edinburgh Math. Soc., **44**, 98-130 (1926)

13. H. von Foerster "Some remarks on changing populations", pp 382-407 in F. Stohlman Jr. (Ed.), "The Kinetics of Cell Proliferation", Grune and Stratton, (1959)

14. W.S.C Gurney, S.P. Blythe and R.M. Nisbet "Nicholson's Blowflies Revisited", Nature(Lond.), **287**, 17-21 (1980)

15. S.P. Blythe, R.M.Nisbet and W.S.C. Gurney "Instability and Complex Dynamic Behaviour in Population Models with Long Time Delays", Theor. Pop. Biol., **22**, 147-176 (1982)

16. N. Ash and B. Greenberg "Developmental Temperature Responses of the Sibling Species *Phaenicia sericata* and *Phaenicia pallescens*", Annals Entomol. Soc. America, **68**, 197-200 (1975)

17. J.H. Lawton "A Population Study on Larvae of the Damselfly *Pyrrhosoma nymphula* (Sulzer) (Odonata : Zygoptera)", Hydrobiologia, **36**, 33-52 (1971) ; and pers. comm. (1982)

18. R.M. Nisbet and W.S.C. Gurney "Modelling Fluctuating Populations", Ch. 4, Wiley, New York (1982)

19. P. Maas, R.M. Nisbet and W.S.C Gurney "SOLVER: an adaptable program template for initial value problem solving", Applied Physics Industrial Consultants, University of Strathclyde, Glasgow, (1982)

S.P. Blythe, R.M. Nisbet and W.S.C. Gurney

Department of Applied Physics,
University of Strathclyde,
John Anderson Building,
107, Rottenrow,
Glasgow G4 ONG

Blood viscosity and its determinants in clinical medicine

C.D. FORBES, G.D.O. LOWE and J.C. BARBENEL

INTRODUCTION

In mortality studies in the Western world the prime cause of death is cardiovascular (about 55 per cent) and the greatest part of this is due to tissue ischaemia resulting from occlusion of arteries by atheromatous deposits or by the formation of thrombus. The organs most commonly at risk are heart, brain, legs and gut if major vessels are primarily affected but a significant mortality and morbidity also exist if small vessels of the microcirculation are affected, for example, in diabetes mellitus.

Ischaemia of tissue is almost always due to some interference with blood supply so that nutrient and oxygen delivery becomes deficient and tissue function declines and death of cells then occurs. The obstruction to the flow of blood is not only anatomical, by atheromatous deposits or thrombus formation but also depends on the fluidity of the blood; the higher the blood viscosity the lower the flow rate.

The purpose of this chapter is to define the terminology used in blood rheology, discuss briefly the methods of measurement of blood viscosity and its determinants and indicate the clinical situations where such measurements may be of clinical significance.

TERMINOLOGY

Rheology deals with the deformation and flow of matter and haemorheology with the flow properties of blood. The most important function is viscosity which is the resistance to flow of a liquid and is due to the internal friction between adjacent layers. In streamlined flow, where adjacent layers move parallel to each other, the velocity difference is a measure of shearing within the flowing liquid and this velocity gradient is termed the shear rate. The unit of shear rate is velocity (distance per unit time) divided by distance, which is expressed as inverse seconds (s^{-1}). The velocity of flow and the shearing within the fluid is produced by a force, the shear stress. The units of shear stress are pascals (Pa) or millipascals (mPa). The viscosity of a fluid may then be defined as the ratio of shear stress to shear rate, expressed as pascal seconds (Pa.s).

$$\text{Viscosity (mPa's)} = \frac{\text{shear stress (mPa)}}{\text{shear rate } (s^{-1})}$$

Newtonian fluids at a constant temperature have a single constant viscosity, as the ratio of shear stress to shear rate is constant and these bear a linear relationship to each other. The problem however is that blood is non-Newtonian in behaviour (Hess 1915) and to define its rheological characteristics, viscosity measurements have to be carried out at a range of constant definable shear rates. In blood, as the shear stress decreases, the shear rate decreases disproportionately, thus the viscosity increases greatly at lower shear rates. At higher shear rates (above $100s^{-1}$) blood does become Newtonian in behaviour.

Because of its non-Newtonian properties blood also possesses a yield stress i.e. a shear stress necessary to produce any shear rate. Blood will therefore stop flowing if the shear stress falls below the yield stress. This force is necessary to disrupt the rouleaux of red cell aggregates present at rest.

DETERMINANTS OF BLOOD VISCOSITY.

Whole blood viscosity is determined by the packed cell volume, plasma viscosity, red cell aggregation, and red cell deformability. The platelets and white cell count have little effect under normal conditions but in certain pathological situations where these are massively elevated they contribute to both whole blood viscosity and capillary plugging.

Blood viscosity may be most easily measured in a rotational viscometer. The blood sample is placed in a sample cup which is maintained at 37°C and a suspended cylindrical bob is lowered into the sample. Either the sample cup or the bob is rotated, with either a constant speed of rotation (shear rate) with measurement of the torque transmitted by the sample to the torsion strip or else a constant torque (shear stress) with measurement of the resultant rate of rotation (shear rate). Such techniques have been recently reviewed by Dormandy (1981).

Packed Cell Volume (PCV)

The volume of packed cells is the most important determinant of whole blood viscosity and accounts for up to 90 per cent of the recorded value. At any shear rate there is a logarithimic increase in blood viscosity with a linear increase in PCV. Even within the normal accepted range of PCV (0.42 - 0.48) there is a significant increase in blood viscosity. When the PCV rises into the range seen in polycythaemia there is an exponential rise in viscosity which is, of course, more pronounced at low shear rates and this viscosity change is also mirrored by a rise in occlusive cardiovascular events (Pearson & Wetherley-Mein, 1979). Within the range of PCV of 0.40-0.50, at a high shear rate ($230s^{-1}$), viscosity increases by approximately 30 per cent whereas at a low shear rate ($0.77s^{-1}$) viscosity increases by approximately 70 per cent. This concept is of major clinical important when considering the need for therapeutic lowering of PCV in affected patients. The PCV is easily measured by rapid centrifugation of a tube of anticoagulated blood in a Hawkesley microcentrifuge.

Plasma Viscosity

Viscosity of plasma is entirely due to the plasma proteins and their concentration. The effect on viscosity by the different plasma proteins increases with molecular size and as the molecular shape departs from the spherical, as indicated by the length/diameter ratio. Thus the large elongated molecule of fibrinogen (MW 340,000, L:D ratio 18:1) exerts a viscosity effect in the plasma equal to half of that of the globulins, although its concentration is only 10 per cent, and similarly globulins (MW 69,000, L:D ratio 4:1) have a greater effect on viscosity than albumin (Harkness and Whittington, 1970) although the globulin concentration is lower. Of the serum globulins, α_2-macroglobulin (MW 820,000) and IgM (MW 1,000,000) have powerful effects: IgA and IgG_3 also have an effect due to their tendency to form complexes (Preston et al 1978). For clinical purposes it is considered that plasma is a Newtonian fluid and the rate of shear of the viscometer is not of major importance. In general, linear increases in plasma protein concentrations produce logarithmic increases in plasma viscosity. These plasma proteins are readily measured by standard biochemical techniques.

Plasma viscosity can be readily measured in a capillary tube viscometer. When a cosntant force, such as gravity is applied, the rate of plasma flow in the tube, at a constant temperature, is inversely related to the viscosity of the plasma. For multiple samples the new Coulter-Harkness viscometer is to be preferred (Harkness 1981).

Red Cell Aggregation.

There are forces acting between red cells which permit their interlocking into three-dimensional networks on cessation of flow. These forces depend in part on characteristics of the red cell membrane and partly on plasma proteins (Chien 1975). It is the large asymetrical plasma proteins - fibrinogen, α_2-macroglobulin and IgM which have the greatest effect (Schmid-Schonbein et al 1973).

The force required to overcome the forces interlocking the red cell aggregates and to restart flow in the blood is the yield stress (Merrill et al 1963). As before the yield stress alters with a rising value of PCV and with a value of PCV from 0.40 rising to 0.50 there is doubling of the yield stress. In addition there is also a positive correlation with the plasma viscosity value (Pearson et al, 1981; Schmid-Schonbein and Reiger,1981).

Red Cell Deformability.

At rest the normal red blood cell is a biconcave disc, but in flowing blood the cell may assume a variety of shapes to allow it to pass through the capillaries of the microcirculation. This ability to deform contributes significantly to the low values of blood viscosity at high shear rates. Deformability depends on the cell content of haemoglobin, the shape of the cell, membrane flexibility and intracellular viscosity. A variety of method of measurement have been suggested and these have been recently extensively reviewed (Barbenel 1981, Working Group on Red Cell Deformability 1981).

The commonest method used involves filtration of washed, diluted cells through polycarbonate filters of pore size 3-5 um. The clinical relevance of such results are not fully understood or documented as yet.

RHEOLOGICAL CHANGES IN PRIMARY DISORDERS OF BLOOD.

Polycythaemia.

Primary proliferative polycythaemia is one of a group of disorders in which there is uncontrolled proliferation of the elements of the bone marrow. There is a high haemoglobin level with an elevated PCV, high white cell count and platelet count. The whole body red cell mass is increased. The importance of this disorder is the increased incidence of thrombotic vascular events which produce morbidity and premature mortality. Over half such cases present with a vascular event, which is usually arterial but up to one quarter of these events are venous thrombosis (Pearson and Wetherley-Mein 1979). The common sites of arterial thrombosis are in the cerebral and coronary arteries (Videbaek 1950, Pearson 1977).

There is a direct correlation of the death rate in polycythaemia with the extent of elevation of the PCV (Pearson and Wetherley Mein 1978), e.g. the incidence of vascular occlusive events rises from 0.2 episodes/10 patient years in the PCV range 0.40-0.44 to 7.5 episodes/ 10 patient years in the range of PCV of 0.60 and above (Pearson 1981). It is of interest that even within the accepted physiological range there is a rise in incidence of events. It is considered that the most important change is in the PCV and that the well documented elevation of the platelet count is of minor importance only (Szur et al 1959, Pearson et al 1981).

Therapy in this condition is directed towards reduction of the elevated PCV. This may be achieved and maintained either by repeated venesection or by the use of radioactive materials (^{32}P) or by cytotoxic chemotherapy. From recent studies of the Polycythaemic Vera Study Group 1981, those patients treated by venesection seem to have a better prognosis because of a reduced incidence of leukaemia and cancer which is found after use of radioactive phosphorus or chlorambucil. Venesection should be carried out regularly to maintain the PCV below 0.45 and it is probably better to remove a small volume of blood routinely (200-300 ml) than take the risk of hypotension-inducing thrombosis as the result of larger venesection. In addition there is evidence that iron deficiency should be avoided as the hypochromic cells are less deformable than normal and their blood viscosity is higher at an equivalent haemoglobin concentration. The therapeutic suggestion is that small amounts of iron should be replaced to avoid this complication (Hutton 1979).

Diseases Associated With Abnormal Plasma Proteins (Paraproteinaemia).

The large molecular weight proteins with symetrical molecules, IgM, IgA and IgG_3 have marked effects on plasma viscosity and red cell aggregation and as a result significantly influence whole blood viscosity. The early associations were with multiple myeloma, a condition of uncontrolled malignant proliferation of plasma cells in marrow (Wintrobe and Buell 1933). Since then a variety of non-malignant conditions have been described in which there is excess production of macroglobulins with a resultant increase in plasma viscosity (Waldenstrom 1949, Jason et al 1970). These syndromes may present with the symptoms and signs associated with hyperviscosity (the hyperviscosity syndrome, H.V.S.) which are similar despite the profound differences in basic pathology (Fahey et al, 1965, Tuddenham et al 1974, Preston et al, 1978). Such patients often complain of lethargy, weakness, easy fatigue and drowsiness. Visual disturbance is a common feature; the patient may complain of haziness or blurring of vision, double vision or blindness. These eye changes are due to stagnation of flow in the retinal vessels which are grossly congested and the retina may become so anoxic that the vessels rupture to produce patches of haemorrhage or they leak to produce pools of exudated protein; both phenomena impair vision. In addition the cerebral circulation seems to be particularly vulnerable and symptoms associated with reduction in flow are common. These include headache, dizziness, vertigo, visual and auditory disturbance and difficulty in balancing. Rarely other major arteries may have impaired circulation with evidence of renal or cardiac impairment or peripheral gangrene.

In these patients the increased serum viscosity is a reflection of the concentration and the intrinsic viscosity of the circulating paraproteins, in particular the physical properties of the IgM molecule which has a molecular weight of 1 million and a high axial length/width ratio. Thus the IgM fraction exerts a greater effect on viscosity thatn the equivalent number of molecules of the lower molecular weight immunoglobulins e.g. IgG has a molecular weight of 160,000: HVS is therefore not usually found until the level of IgG has risen to 80g/litre. IgA and IgG_3 have greater effects on viscosity because of their tendency to form protein aggregates.

The diagnosis of HVS is easily confirmed by direct measurement and this should be done routinely in patients with myeloma even if symptoms are absent at that time. Treatment can be rapid and effective. Plasmapheresis, a process by which the plasma of the patient is removed and replaced by a normal human plasma solution, results in rapid clinical improvement. However as the abnormal proteins are continuously synthesised the procedure may require to be repeated at intervals life-long or until other types of chemotherapy or radiotherapy can be given to suppress the cells producing the abnormal proteins.

Leukaemia.

Symptoms and signs similar to those described above in HVS may also be found in patients with a variety of leukaemias. This is probably due to the massive increase in the whole blood viscosity resulting from gross elevation of the white cell count (Preston et al 1978). In addition the picture is probably compounded by the presence of leukaemic infiltrates into the brain substance and electrolyte imbalance.

Normal white cells are less deformable than red cells (Bagge et al 1981) and this applies particularly to the leukaemic cell which has a large non-deformable nucleus (Lichtman, 1973, Bagge 1976). In post-mortem samples of tissue from patients dying with leukaemia blockage of small vessels by white cells is commonly seen - leukostasis. This is particularly true of myeloid leukaemia (McKee and Collins 1974) as the myeloblast is significantly less deformable than other white cells.

The incidence of leukostasis increases with the peripheral white cell count and is found in most patients with a count above 200 x 10^9/1 (McKee and Collins 1974). Part of the tendency to a dangerously high level of whole blood viscosity is however obviated in leukaemia by an inverse fall in the packed cell volume as the normal red cell precursors are replaced in the marrow. However this does not always hold and white cell counts over 100 x 10^9/1 with a haemoglobin level of 10g/dl carries a significant mortality from HVS (Harris et al,1978). This phenomenon also highlights the importance of restraining from transfusion of whole blood in such patients until the white cell count is controlled.

If the clinical problem is acute, leucopheresis, a process by which the white cells are removed by continuous centrifugation and the white cell free blood is returned to the patient, may prove life-saving. Such a procedure produces dramatic reversal of symptoms and allows time for conventional cytotoxic therapy to exert its action.

Sickle Cell Disease.

This is one of a series of diseases in which there is a genetic abnormality in the haemoglobin molecule. As a result of a variety of stimuli the haemoglobin molecule may crystallise and the result is a change in shape of the red cell, to a rigid, sickle shape which causes vascular obstruction in small arterioles and capillaries. The stimuli to sickling in patients are anoxia and changes in P_H of the blood but often such a stimulus is apparently absent in vivo. In addition vascular occlusion is compounded by platelet-fibrin thrombosis.

In the laboratory whole blood viscosity from an asymptomatic homozygote when deoxygenated will increase progressively as the oxygen saturation falls and less deformable, sickled cells are formed. This is more pronounced at higher shear rates where deformability is of paramount importance (Usami et al 1975). Such sickle changes in the red cell are associated with profound metabolic changes in the cell - in particular reduced level of ATP and potassium and an increased content of calcium (Eaton et al 1979). The water content and the osmality are decreased and there is an increase in the mean corpuscular haemoglobin concentration and hence in the internal viscosity of the cell.

In addition to the obvious cellular abnormality there is an increase in plasma viscosity which is due mainly to increased globulins as the fibrinogen level in the steady state is not elevated (Chien et al 1976, Richardson et al 1979). During an acute crisis there is a rapid acute phase rise in fibrinogen which significantly increases the plasma viscosity as well as increasing the tendency for red cell aggregation.

RHEOLOGICAL CHANGES IN ARTERIAL OCCLUSION

Epidemiological Studies.

No studies have been carried out of measurement of blood viscosity prospectively in arterial diseases. However, incidentally, several studies have measured the most important determinants, the PCV (or haemoglobin level) and also the plasma fibrinogen level (or the erythrocyte sedimentation rate).

There is now evidence from such studies that even within the normal range of PCV there is a positive correlation of the incidence of occlusive arterial events with a rising PCV. High normal haemoglobin levels have been shown to be subsequently associated with cerebral infarction (Kannel et al 1972) and myocardial infarction (Bottiger and Carlson 1973, Elwood et al 1974, Dawber et al 1977, Sorlie et al 1981, Baker et al 1982). Similarly elevation of the fibrinogen levels are associated with subsequent death from coronary artery disease (Bottiger and Carlson 1973, Meade et al 1980).

Risk Associations.

It is well known that from the clinical point of view the following risk associations are important in the onset of arterial disease. These are age, sex, cigarette smoking, diabetes mellitus, blood pressure and blood lipids.

Age. The incidence of arterial thrombosis is age dependent and rises sharply after the age of 40 years. Numerous studies have shown an association of a rising plasma fibrinogen with advancing age and this is reflected in the ESR which also rises with age (Bottiger and Svedberg 1967). Our own studies confirm this observation and have also shown a similar rise in plasma viscosity and in whole blood viscosity (corrected to a standard PCV of 0.45) Lowe et al 1980b). The packed cell volume changes little with age during adult life but there is said to be a reduction in red cell deformability (Reid et al 1976).

Sex. Men are more liable to arterial thrombosis than women, however much of this difference is accounted for by the significant protection women seem to have before and 10 years after the menopause. This may be partly explained by the lower packed cell volume in menstruating women and the chronic iron deficient state associated with menstruation. Plasma viscosity is also lower in pre-menopausal women than men (Mayer 1964). There appear to be no significant differences in plasma fibrinogen or in red-cell deformability (Drummond et al 1979).

Cigarette smoking. This is thought to be the most important risk association for arterial disease from epidemiological studies.

Blood viscosity, packed cell volume, fibrinogen and plasma viscosity are higher in men examined over a wide range of age (Dintenfass 1975, Lowe et al 1980b) and this is particularly so for the young smoker who has a "rheological profile" of a non-smoking middle-aged man. Part of these findings can be explained by the elevated packed cell volume associated with the anoxia due to carbon monoxide inhalation but the cause of the elevated fibrinogen is unknown. Using a filtration technique with washed red cells no changes in red cell deformability have been found.

Diabetes mellitus. Arterial degeneration with atheromatous deposition in large vessels and injury to the small vessels of the microcirculation is the commonest and as yet unsolved complication of diabetes mellitus (Jarrett and Keen 1975, Kohner and Oakley 1975).

In a series of independent studies in well controlled diabetes of both insulin dependent and mature onset there were significantly higher values of whole blood viscosity, (also corrected to a PCV of 0.45), plasma viscosity and fibrinogen (Hoare, Barnes and Dormandy 1976, Barnes et al 1977, Lowe et al 1980d). In addition there is evidence that red cell deformability is also greater in diabetes using different test systems (Schmid-Schonbein and Volger 1976, McMillan et al 1978).

There is now good evidence that some of these rheological abnormalities will return to normal with better control as measured by a decline in the levels of glucosylated proteins, in particular haemoglobin A_{ic} (Barnes, 1981, Leiper et al 1982). As yet no long term studies have been carried out which enable us to say that better control and hence more normal rheological factors reduces the incidence of vascular occlusive events or disease of the microcirculation.

Blood pressure. Elevation of blood pressure, in particular the systolic level carries a significant risk of increased morbidity and mortality from vascular events. There is evidence of abnormal rheological factors, in particular, elevation of PCV, plasma viscosity and fibrinogen (Tibblin et al 1966, Isles et al 1981, Dintenfass 1981). Epidemiological studies also support these findings with positive correlations being found between packed cell volume or haemoglobin level and both systolic and diastolic blood pressure (McDonagh et al 1965, Kannell et al 1972). Also significantly increased levels of fibrinogen are found in malignant hypertension (Gavras et al 1975), and a positive correlation has been found between fibrinogen levels and blood pressure in a population study (Korsan-Bengtsen et al 1972).

Lipids. There is clear evidence of an association between abnormalitieies of blood lipids and the onset of premature arterial disease due to atheroma. In patients with genetic or acquired lipid abnormalities such effects are more marked e.g. an untreated homozygote with type II hyperlipoproteinaemia may have myocardial ischaemia in his early teens and die in his early twenties of myocardial infarction.

There seems to be no close association of rheological factors and hyperlipaemia after meals (Charm et al 1963).

However we have previously shown gross elevation of the fibrinogen levels in patients with type II hyperlipoproteinaemia (Lowe et al 1979) and epidemiological studies show a positive correlation, within the normal ranges, of cholesterol with fibrinogen (Karsar-Bergsten et al 1972) and packed cell volume (McDonagh et al 1965, Elwood et al 1970).

Hyperviscosity in Chronic Occlusive Arterial Disease.
Alteration of blood viscosity and its components is well recognised in coronary artery disease (Mayer 1964, Nicolaides et al 1977, Lowe et al 1980a) and in peripheral arterial disease (Dormandy et al 1973a, Stormer et al 1974, Blunt et al 1980). The prime factors responsible are elevation of the PCV and also fibrinogen. The extent of the viscosity abnormality correlates with the prognosis for the limb in cases of peripheral arterial disease: a poor prognosis being associated with increased blood viscosity (Dormandy et al 1973b), elevatio of haemoglobin level (Bouhoutsos et al 1974; Bailey et al 1979) and with fibrinogen excess (Dormandy et al 1973b; Hamer et al 1973, Harris et al 1978). We have previously shown a positive correlation between blood viscosity, PCV and fibrinogen and the extent of coronary artery disease demonstrated angiographically (Lowe et al 1980b) and previous workers have found a positive correlation between the PCV and the extent of cerebral atherosclerosis found at autopsy (Tohgi et al 1978).

Blood Viscosity and Venous Thrombosis.
Flow conditions in the venous side of the circulation has different implications compared to the arterial tree. Low flow and low shear conditions exist in the deep veins of the legs particularly in patients immobilised by acute stroke or confined to bed after surgery, during general anaesthesia and in patients with severe heart failure. It is in the pockets of the valves of the

veins that flow may become stagnant and it is here tha thrombi originate as activated clotting factors accumulate and trigger fibrin formation.

Many of the clinical risk associations for the development of venous thrombosis are also associated with abnormalities of blood viscosity. These include pregnancy and oestrogen therapy (Buchan and MacDonald, 1981, Lowe et al 1980 (c)) malignancy (Tietjen et al,1977) surgery (Harvey-Kemble and Hickman 1972, Lowe et al 1981) heart failure (Eisenberg 1964) polycythaemia (Pearson and Wetherley-Mein 1978), obesity (McDonough et al 1965, Bennett et al 1966) and age (Bottiger and Svedberg 1967, Lowe et al 1980 (b)).

Perhaps the major interest in venous thrombosis and rheological factors is the possibility of altering the natural history of the disease by changing some, or all of such factors. In a previously reported study we successfully reduced the incidence and extent of post-operative venous thrombosis in patients operated on for fractures of the femoral neck. In this study we defibrinogenated the patients with the snake venon, ancrod, and maintained the low levels of fibrinogen for the whole of the post-operative period (Lowe et al 1978). Similar results have been shown in femoral vein grafts in dogs (Ishimaru et al 1979) and following pulmonary embolism in dogs (Meissner et al 1976). Further studies in prevention of venous thombosis with reduction of both fibrinogen and packed cell volume are in progress in our group.

Conclusions.

The relationship of blood viscosity and its determinants with vascular disease is only now becoming obvious as the technology to make viscosity measurements becomes available. However some of these are still primitive especially the methods available for measurement of red cell deformability. Results, therefore, have to be treated with some reservation until better measurement techniques are available. Also it is not really clear whether the well documented changes in vascular disease are the result, the cause, or just coincidental observations due to injury to tissues.

In some ways events have overtaken these measurements as various procedures have become available to alter blood viscosity either in the long or short-term. Currently studies are in progress in most of the conditions described using techniques such as long term venesection, haemodilution, defibrinogenation and fibrinogen lowering agents. The results of placebo controlled double blind studies will give us the answers to some of these questions.

REFERENCES.

1. Bagge, U. (1976). Granulocyte rheology. Blood Cells. 2, 481-490.

2. Bagge, U., Branemark, P.I., and Skalak, R.(1981). Measurement and influence of white cell deformability. In: Clinical Aspects of Blood Viscosity and Cell Deformability. G.D.O. Lowe, J.C. Barbenel, C.D. Forbes (Eds.) Springer-Verlag, New York. pp. 27-36.

3. Bailey, M.J., Yates, C.J.P., Johnston, C.L.W., Somerville, P.G. and Dormandy, J.A. (1979). Pre-operative haemoglobin as a predictor of outcome of diabetic amputations. Lancet ii, 168-170.

4. Baker, I.A., Eastham, R., Elwood, P.C., Etherington, M., O'Brien, J.R., Sweetnam, P.M. (1982). Haemostatic factors associated with ischaemic heart disease in men aged 45-64 years. The Speedwell Study. British Heart Journal, 47, 490-494.

5. Barbenel, J.C.(1981). The measurement of red blood cell deformability. In: Clinical Aspects of Blood Viscosity and Cell Deformability. G.D.O. Lowe, J.C. Barbenel, C.D. Forbes (Eds.) Springer-Verlag, New York, pp. 37-45.

6. Barnes, A.J. (1981). Blood viscosity in diabetes mellitus. In: Clinical Aspects of Blood Viscosity and Cell Deformability. G.D.O. Lowe, J.C. Barbenel, C.D. Forbes (Eds.) Springer-Verlag, New York, pp. 151-162.

7. Barnes, A.J., Locke, P., Scudder, P.R., Dormandy, T.L., Dormandy, J.A. and Slack, J. (1977). Is hyperviscosity a treatable component of diabetic microcirculatory disease? Lancet, ii, 789-791.

8. Bennett, N.B., Ogston, C.M., McAndrew, G.M. and Ogston, D. (1966). Studies on the fibrinolytic enzyme system in obesity. Journal of Clinical Pathology, 19, 241-243.

9. Blunt, R.G., George, A.J., Harlow, R.A., Strachan, C.J. and Stuart, J. (1980). Hyperviscosity and thrombotic changes in idiopathic and secondary Raynauds syndrome. British Journal of Haematology, 45, 651-658.

10. Bottiger, L.E. and Carlson, L.A. (1973). The Stockholm Prospective Study (2). New events of coronary heart disease in man in relation to findings at initial examination. 9 year follow-up. In: Early Phases of Coronary Heart Disease. Waldestrom, J., Larston, T., and Ljungstedt, T. (Eds.) Nordiska Bokhandelns Forlag, Stockholm. pp 158-179.

11. Bottiger, L.E., and Svedberg, C.A. (1967). Normal erythrocyte sedimentation rate and age. British Medical Journal, 2, 85-87.

12. Bouhoutsos, J., Morris, T., Chavatzas, D and Martin, P. (1974). The influence of haemoglobin and platelet levels on the results of arterial surgery. British Journal of Surgery, 51, 984-986.

13. Buchanan, P.C. and MacDonald, H.N. (1980). Altered haemorheology in oral contraceptive users. British Medical Journal, i, 978-979.

14. Charm, S.E., McComis, W., Tejada, C., Kurland, G. (1963). Effect of a fatty meal on whole blood and plasma viscosity. Journal of Applied Physiology, 18, 1217-1220.

15. Chien, S.(1975). Biophysical behaviour of red cells in suspensions. In: The Red Blood Cells. Vol. 2. Second Edition. D.M. Surgenor (Ed.) Academic Press, London, pp 1031-1133.

16. Chien, S., Usami, S., Jan K.M., Smith, J.A., Bertles, J.F. (1976). Blood rheology in sickle cell disease. In: Proceedings of the Symposium on Molecular and Cellular Aspects of Sickle Cell Disease. Hercules, J.I., Cottam, G.L., Waterman, M.R., Schecter, A.N. (Eds.) U.S. Dept. of Health Education and Welfare, Bethesda. Publication No. (NIH) 76-1007. pp.277-300.

17. Dawber, T.R., Wolf, P.A.,Cotton, T. and Nickerson, R.J. (1977). Risk Factors in Brain and Heart Infarct. Zaloh, K.G., Kaufmann, W., Hassman, K.A. and Hassman, V. (Eds.).Springer-Verlag, Berlin, pp 226-252.

18. Dintenfass, L..(1975). Elevation of blood viscosity, aggregation of red cells, haematocrit values and fibrinogen levels in cigarette smokers. Medical Journal of Australia, 1, 617-620.

19. Dintenfass, L. (1981). Hyperviscosity in Hypertension. Pergamon Press, Oxford.

20. Dormandy, J.A. (1981). Measurement of Whole-blood viscosity. In: Clinical Aspects of Blood Viscosity and Cell Deformability. G.D.O. Lowe, J.C. Barbenel, C.D. Forbes (Eds.). Springer-Verlag, New York, pp.67-78.

21. Dormandy, J.A., Hoare, E., Colley, J., Arrowsmith, D.E., Dormandy, T.D. (1973). Clinical, haemodynamic, rheological, and biochemical findings in 126 patients with intermittent claudication. British Medical Journal, iv, 576-581.

22. Dormandy, J.A., Hoare, E., Khattab, A.M., Arrowsmith, D.E. and Dormandy, T.L. (1973). Prognostic significance of rheological and biochemical findings in patients with intermittent claudication. British Medical Journal, iv, 581-583.

23. Drummond, M.M., Lowe,G.D.O., Belch, J.J.F., Forbes, C.D. and Barbenel, J.C. (1979). An assessment of red cell deformability using a simple filtration method. Thrombosis and Haemostasis 42, 106-107.

24. Eaton, J.W., Jacob, H.S. and White, J.G. (1979). Membrane abnormalities in irreversibly sickled cells. Seminars of Haematology, 16, 52-61.

25. Eisenberg, S.(1964). Changes in blood viscosity haematocrit value and fibrinogen concentration in subjects with congestive cardiac failure. Circulation, 30, 686-693.

26. Elwood, P.C., Mahler, R., Sweetnam, P.M., Moore, F., Welsley, E. (1970). Association between circulating haemoglobin level, serum cholesterol and blood pressure. Lancet 1, 589-591.

27. Elwood, P., Water, W.E., Benjamin, I.T., and Sweetnam, P.M. (1974). Mortality and anaemia in women. Lancet, 1, 891-894.

28. Fahey, J.L., Barth, W.F. and Solomon, A. (1965). Serum hyperviscosity syndrome. Journal of the American Medical Association, 192, 464-467.

29. Gavras, H., Oliver, N., Aitchison, J., Begg, C., Briggs, J.D., Brown, J.J., Horton, P.W., Lee, F., Lever, A.F., Prentice, C., Robertson, J.I.S. (1975). Abnormalities of coagulation and the development of malignant phase hypertension. Kidney International 8, S-252-S261.

30. Hamer, J.D., Ashton, F. and Meynell, M.J. (1973). Factors influencing prognosis in the surgery of peripheral vascular disease: platelet and adhesiveness, plasma fibrinogen and fibrinolysis. British Journal of Surgery, 60, 386-389.

31. Harkness, J. (1981). Measurement of plasma viscosity. In: Clinical Aspects of Blood Viscosity and Cell Deformability. G.D.O. Lowe, J.C. Barbenel, C.D. Forbes, (Eds.) Springer-Verlag, New York, pp. 79-87.

32. Harkness, J. and Whittington, R.B.(1970). The viscosity of human blood plasma: its measurement in health and disease. Biorheology, 8, 171-193.

33. Harris, J.W., Brewster, H.H., Ham, T.H., Castle, W.B. (1956). Studies on the destruction of red cells. The biophysics and the biology of sickle-cell disease. Archives of Internal Medicine, 97, 145-168.

34. Harris, P.L., Harvey, D.R. and Bliss, B.P.(1978). The importance of plasma lipids, glucose, insulin and fibrinogen in femoropopliteal surgery. British Journal of Surgery, 65, 197-200.

35. Harvey-Kemble, J.V. and Hickman, J.A. (1972). Post-operative changes in blood viscosity and the influence of haematocrit and plasma fibrinogen. British Journal of Surgery, 59, 629-632.

36. Hess, W. (1915). Gehorcht das Blut dem allgeimine Stromugsgesety der Flussigkeiten. Arch. Gesampt Physiol. 162, 187-197.

37. Hoare, E.M., Barnes, A.J. and Dormandy, J.A. (1976). Abnormal blood viscosity in diabetes mellitus and retinopathy. Biorheology, 13, 21-25.

38. Hutton, R.D. (1979). The effect of iron deficiency on whole blood viscosity in polycythaemia patients. British Journal of Haematology, 43, 191-199.

39. Ishimaru, S., Hirayama, T., Furakawa, K. and Takahashi, M. (1979). Antithrombotic effect of defibrinogenating enzyme on experimental small calibre vein replacement using PTFE graft. Thrombosis and Haemostasis, 42, p.231.

40. Isles, C., Lowe, G.D.O., Drummond, M.M., Forbes, C.D., Kennedy A.C., and Lever A.F. (1981). Blood rheology in malignant phase hypertension. Scandinavian Journal of Clinical and Laboratory Investigation. 41, Supplement 156, p.171.

41. Jarrett, R.J., and Keen H. (1975). Diabetes and atherosclerosis. In: Keen, H., and Jarrett, R.J. (Eds.) Complications of diabetes. Edward Arnold, London. pp.179-203.

42. Jasin, H.E., LoSpalluto, J. and Ziff, M. (1970). Rheumatoid hyperviscosity syndrome. American Journal of Medicine, 49, 484-493.

43. Kamel, W.B., Gordon, T., Wolf, P.A. and McNamara, P. (1972). Haemoglobin and the risk of cerebral infarction: the Framingham Study. Stroke, 3, 409-420.

44. Kohner, E.M., and Oakley, N.W. (1975). Diabetic retinopathy. Metabolism, 25, 1085-1102.

45. Korsan-Bengtsen, K., Wilhehnsen, L., Tibblin, G. (1972). Blood coagulation and fibrinolysis in a random sample of 788 men 54 years old. II relations of the variables to risk factors for myocardial infarction. Thrombosis et Diathesis Haemorrhagica, 28, 99-108.

46. Leiper, J.H., Lowe, G.D.O., Anderson, J.,Burns, P., MacCuish, A.C., Manderson, W.G., Forbes, C.D. and Barbenel, J.C. (1982). Effects of diabetic control and biosynthetic human insulin on blood rheology in established diabetics (in press.).

47. Lichtman, M.A. (1973). Rheology of leukocytes, leukocyte suspensions and blood in leukaemia. Journal of Clinical Investigation, 52, 350-358.

48. Lowe, G.D.O., Campbell, A.F., Meek, D.R., Forbes, C.D.,Prentice, C.R.M., Cummings, S. (1978). Subcutaneous ancrod in prevention of deep vein thrombosis after operation for fractured neck of femur. Lancet, ii, 698-700.

49. Lowe, G.D.O., Drummond,M.M., Third, J.L.H.C., Bremner, W.F.,Forbes, C.D., Prentice, C.R.M. and Lawrie, T.D.V. (1979). Increased plasma fibrinogen and platelet-aggregates in type II hyperlipoproteinaemia. Thrombosis and Haemostasis, 42, 1503-1507.

50. Lowe, G.D.O., Drummond,M.M., Lorimer, A.R., Hutton, I., Forbes, C.D., Prentice, C.R.M. and Barbenel, J.C. (1980a). Relation between the extent of coronary artery disease and blood viscosity. British Medical Journal, i, 573-674.

51. Lowe, G.D.O., Drummond, M.M., Forbes, C.D., Barbenel, J.C. (1980b). The effects of age and cigarette smoking on blood and plasma viscosity in men. Scottish Medical Journal, 25, 13-17.

52. Lowe, G.D.O., Drummond, M.M., Forbes, C.D. and Barbenel, J.C. (1980d). Increased blood viscosity in young women using oral contraceptives. American Journal of Obstetrics and Gynaecology. 137, 840-842.

53. Lowe, G.D.O., Lowe, J.M., Drummond, M.M., Reith, S., Belch, J.J.F., Kesson, C.M., Wylie, A., Foulds, W.S., Forbes, C.D., MacCuish, A.C., and Manderson, W.G. (1980d). Blood viscosity in young male diabetics with and without retinopathy. Diabetologica, 18, 359-363.

54. Lowe, G.D.O., Drummond, M.M., Forbes, C.D., Prentice, C.R.M. and Barbenel, J.C. (1981) Blood and plasma viscosity in prediction of venous thrombosis. Scandinavian Journal of Clinical and Laboratory Investigation. 41 (Supplement 156). 307.

55. McDonough, J.R., Hames, C.G.,Garrison, G.E., Stubb, S.C., Lichtman, M.A., Hegelfinger, D.C. (1965). The relationship of the haematocrit to cardiovascular states of health in the negro and white population of Evans County, Georgia. Journal of Chronic Disease, 18, 243-257.

56. McKee, L.C., Collins, R.D. (1974). Intravascular leukocyte thrombi and aggregates as a cause of morbidity and mortality in leukaemia. Medicine, (Baltimore), 53, 463-478.

57. McMillan, D.E., Utterback, N.G., LaPuma, J.(1978). Reduced erythrocyte deformability in diabetes. Diabetes, 27, 895-901.

58. Mayer, G.A. (1964). Blood viscosity in healthy subjects and patients with coronary heart disease. Canadian Medical Association Journal, 91, 951-954.

59. Meade, T.W., North, W.R.S. ,Chakrabarti, R., Haines, A.P., Stirling, Y. (1977). Population-based distributions of haemostatic variables. In: Haemostasis. D. Thomas (Ed.) British Medical Bulletin, 33, 283-288.

60. Meade, T.W., North, W.R.S., Chakrabarti, R., Stirling, Y., Haines, A.P., Thompson, S.G. and Brozovic, M. (1980). Haemostatic function and cardiovascular death: early results of a prospective study. Lancet, i, 1050-1054.

61. Meissner, A.G., O'Sullivan, G. and MacBeth, R.A. (1976). The effect of defibrinogenation on pulmonary embolism in dogs. Thrombosis et Diathesis Haemorrhagica 36, 140-147.

62. Merrill, E.W., Gilliland, G.R., Cokelet, G., Shin, H., Britten, A., Wells, R.E. (1963). Rheology of human blood, new and at zero flow. Effects of temperature and haematocrit level. Biophysical Journal 3, 199-213.

63. Nicolaides, A.N., Bowers, R., Horbourne, T., Kidne P.H., and Besterman, E.M. (1977). Blood viscosity red-cell flexibility, haematocrit and plasma fibrinogen in patients with angina. Lancet ii, 943-945.

64. Pearson, T.C., Humphrey, P.R.D., Thomas, D.J. and Wetherley-Mein, G. (1981). Haematocrit and blood viscosity, cerebral blood flow and vascular occlusion. In:Clinical Aspects of Blood Viscosity and Cell Deformabilty. G.D.O. Lowe, J.C.Barbenel, C.D. Forbes (Eds.). Springer-Verlag, New York. pp.97-107.

65. Pearson, T.C. and Wetherley-Mein, G. (1978). Vascular occlusive episodes and venous haematocrit in primary proliferative polycythaemia. Lancet, ii, 1219-1222.

66. Pearson, T.C. and Wetherley-Mein, G. (1979). The cause and complications of idiopathic erythocytosis. Journal of Clinical and Laboratory Haematology, 1, 189-196.

67. Polycythaemia Vera Study Group (1981). Increased incidence of acute leukaemia in polycythaemia vera associated with chlorambucil therapy. New England Journal of Medicine, 304, 441-447.

68. Preston, F.E., Cooke, K.B., Foster, M.E., Winfield, D.A., and Lee, D. (1978). Myelomatosis and the hyperviscosity syndrome. British Journal of Haematology, 38, 517-530.

69. Preston, F.E., Sokol, R.J., Lilleyman, J.S., Winfield, D.A., and Blackburn, E.K. (1978). Cellular hyperviscosity as a cause of neurological symptoms in leukaemia. British Medical Journal, 1, 476-478.

70. Reid, H.L., Barnes, A.J., Lock, P.J.,Dormandy, J.A., Dormandy, T.L. (1976). A simple method for measuring erythrocyte deformability. Journal of Clinical Pathology, 28, 855-878.

71. Richardson, S.G.N., Matthews, K.B., Stuart, J., Geddes, A.M., Wilcox, R.M. (1979). Serial changes in coagulation and viscosity during sickle-cell crises. British Journal of Haematology, 41, 95-103.

72. Schmid-Schonbein, H., Gallasch, G., Volger, E., Klose, H.J. (1973). Microrheology and protein chemistry of pathological red cell aggregation. (blood sludge) studied in vitro. Biorheology,10, 213-227.

73. Schmid-Schonbein, H. and Reiger, H. (1981). Usovolaemic haemodilution. In: Clinical Aspects of Blood Viscosity and Cell Deformability. G.D.O. Lowe, J.C.Barbenel, C.D. Forbes, (Eds.) Springer Verlag, New York. pp. 221-226.

74. Schmid-Schonbein, H., Volger, E. (1976). Red cell aggregation and red cell deformability in diabetes. Diabetes 25 (Suppl.2) 897-902.

75. Sorlie, P.D., Garcia-Palmieri, M .R., Costos, R., Havlik, R.J. (1981). Haematocrit and risk of coronary heart disease: the Puerto Rico Heart Health Program. American Heart Journal, 101, 456-461.

76. Stormer, B., Horsch, R., Kleinschmidt, F., Loose, D., Bruster, H. and Kremer, K. (1974). Blood viscosity in patients with peripheral vascular diseases in the areas of low shear rates. Journal of Cardiovascular Surgery, 15, 577-584.

77. Szur, L., Lewis, S.M., Goolden, A.W.G.(1959). Polycythaemia vera and its treatment with radioactive phosphorus. Quarterly Journal of Medicine, 29, 397-424.

78. Tibblin, G., Bergent, S.E., Bjure, J., Wilhelmsen, L. (1966). Haematocrit, plasma protein, plasma volume and viscosity in early hypertensive disease. American Heart Journal, 72, 165-176.

79. Tietjen, G.W., Chien, S., Scholz, P., Gump,F.E., and Kinney, J.M. (1977). Changes in blood viscosity and plasma proteins in carcinoma. Journal of Surgical Oncology, 9, 53-59.

80. Tohgi, H., Yamanouchi, H., Murakami, M. and Kameyama, M. (1978). Importance of the haematocrit as a risk factor in cerebral infarction. Stroke, 9, 369-374.

81. Tuddenham,E.G.D., Whittaker, J.A., Bradley, J., Lilleyman, J.S., and Jones, D.R. (1974). Hyperviscosity syndrome in IgA multiple myeloma. British Journal of Haematology 27, 57-76.

82. Usami, S., Chien, S., Scholz, P.M., Bertles, J.F. (1975). Effect of deoxygenation on blood rheology in sickle cell disease. Microvascular Research, 9, 324-334.

83. Videback, A. (1950). Polycythaemia vera: cause and prognosis. Acta Medica Scandinavica 138, 179-187.

84. Waldenstrom, J. (1944). Incipient myelomatosis or essential hyperglobulinaemia with fibrinogenopenia - a new syndrome. Acta Medica Scandinavica, 117, 216-247.

85. Wintrobe, M.M., Buell, M.V. (1933). Hyperproteinaemia associated with multiple myeloma. Bulletin of the John Hopkins Hospital, 52, 156-165.

86. Working Group Report on Red Cell Deformability. Methods and Terminology. Proceedings of the Second European Conference on Clinical Haemorheology. September, 1981.

Numerical studies of reaction-diffusion problems

V.S. MANORANJAN

1. INTRODUCTION

In this paper we report some of the numerical results obtained for certain Reaction-Diffusion equations such as the Fisher's equation,

$$u_t = u_{xx} + f(u) \tag{1}$$

which appears in population genetics, and the Fitzhugh-Nagumo system

$$\begin{aligned} u_t &= u_{xx} + u(1-u)(u-a) - v \\ v_t &= b(u-dv) \qquad 0 < a < \tfrac{1}{2} \\ & \qquad\qquad\qquad b > 0,\ d \geq 0 \end{aligned} \tag{2}$$

which governs the conduction of electrical impulses in a nerve axon.

These equations have been studied extensively by many authors for both pure initial and initial boundary value problems and the interested reader is referred to [2,3,4,5,6]. The numerical method employed in solving the respective problems to be discussed below, is a finite element Galerkin method and the reader is referred to [9] for a full description of this method.

2. FISHER'S EQUATION

a) Let $f(u) = u(1-u)$ in (1) and we shall look at the pure initial value problem with

$$u(x,0) = u_o(x) \quad , \quad -\infty < x < \infty \ .$$

It was shown by Kolmogorov et al [6] that the solution $u(x,t)$ approaches a travelling front profile of speed 2 as $t \to \infty$, if

$$u_o(x) = \begin{cases} 1 \ , & x < 0 \\ 0 \ , & x > 0 \ . \end{cases} \tag{3}$$

Later McKean [8] proved that if $u_o(x)$ is any non-negative continuous monotonic function such that

$$\lim_{x \to -\infty} u_o(x) = \lambda e^{\beta x} \quad , \quad \lambda > 0$$

and (4)

$$u_o(+\infty) = 1$$

then there exists a wave solution of asymptotic speed

$$c = \beta + \frac{1}{\beta} \quad , \quad \text{when} \quad 0 < \beta \leq 1 \ . \tag{5}$$

Our numerical results show that the speeds of the travelling fronts converge to the value

$$c = 2 \quad , \quad \text{when} \quad \beta > 1 \quad \text{(see Table (1))} \ .$$

$\beta(> 1)$	c
$^3/\sqrt{6}$	1.9963
1.4	1.9949
2.5	1.9938
5.0	1.9937

$$-200 \leq x \leq 100 \quad , \quad 0 \leq t \leq 50$$

$$u_o(x) = \begin{cases} \frac{1}{2} e^{\beta x} \ , & x \leq 0 \\ 1 - \frac{1}{2} e^{-\beta x} \ , & x > 0 \end{cases} \tag{6}$$

Table (1)

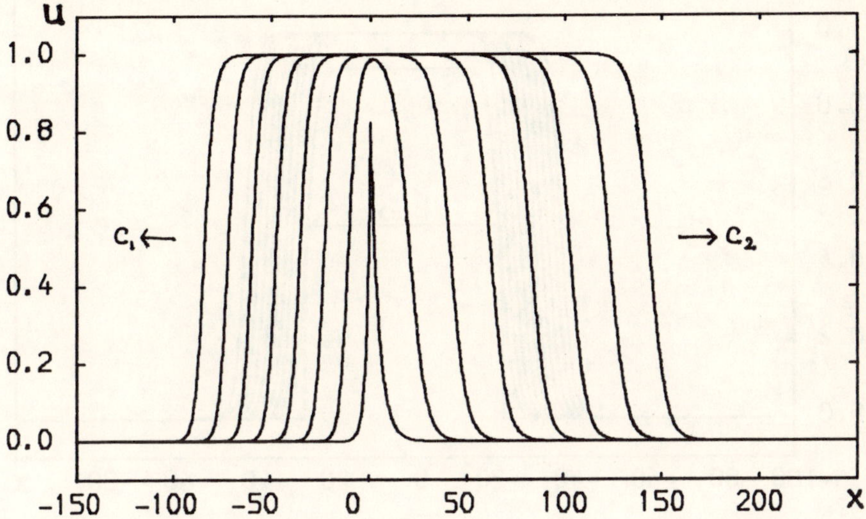

$$u_0(x) = \begin{cases} \frac{1}{2}\exp\{\frac{1}{2}(x+1)\} & , \quad x \le 0 \\ \frac{1}{2}\exp\{-\frac{1}{4}(x-2)\} & , \quad x > 0 \end{cases}$$

$$c_1 \sim 2.5 \quad , \quad c_2 \sim 4.25$$

Fig.(1) Fronts with different speeds.

These results agree with those of Rothe [10], Larson [7], and Uchiyama [13].

An exact solution found recently by Ablowitz and Zeppetella [1] for a special wave speed $c = {}^5/\sqrt{6}$ which can be written as

$$u = {}^1/\{1 + r\exp({}^{-\xi}/\sqrt{6})\}^2 \tag{7}$$

$$\xi = x + {}^5/\sqrt{6}\ t \quad , \quad r \text{ a positive parameter} ,$$

is also consistent with (4) and (5) , since as $x \to -\infty$, $u \sim \frac{1}{r^2}\exp({}^{-2\xi}/\sqrt{6})$, with $\beta = {}^2/\sqrt{6}$ to give $c = \beta + \frac{1}{\beta} = {}^5/\sqrt{6}$. One of the particularly interesting

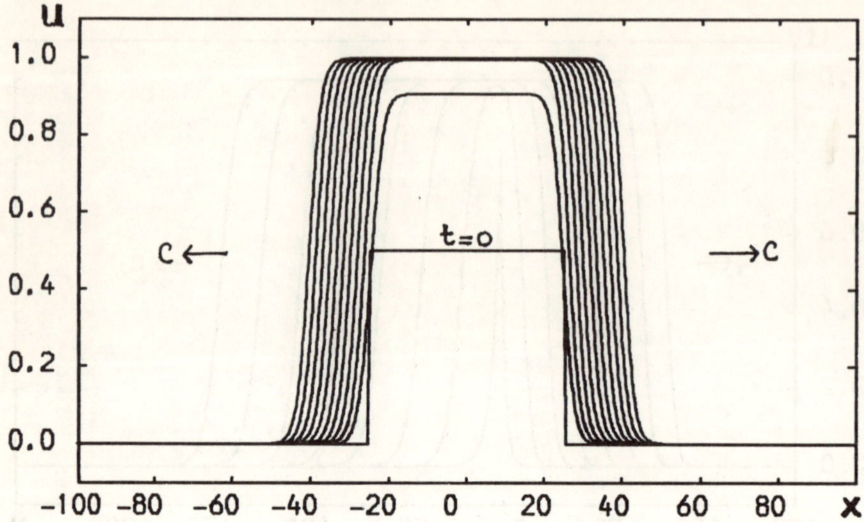

$$u_0(x) = \begin{cases} 0.5(> a=0.25) & , \quad -25 \le x \le 25 \\ 0 & , \quad \text{otherwise.} \end{cases}$$

$c = 0.3536$ ($\sim$ Huxley speed)

Fig.(2) Formation of Huxley fronts.

results is that we can obtain fronts travelling in opposite directions with different speeds depending on the behaviour of $u_o(x)$ at both plus and minus infinity. (Fig.(1)).

b) When $f(u) = u(1-u)(u-a)$, $0 < a < \frac{1}{2}$ in (1) , the equation is known as the degenerate Fitzhugh-Nagumo equation and the pure initial value problem is known to have the so called Huxley solution

$$u = \{1 + \exp(^{-\xi}/\sqrt{2})\}^{-1}$$
$$\xi = x + ct \; , \; c = \sqrt{2}(\tfrac{1}{2}-a) \; . \tag{8}$$

Also, this equation is known to exhibit a threshold property,

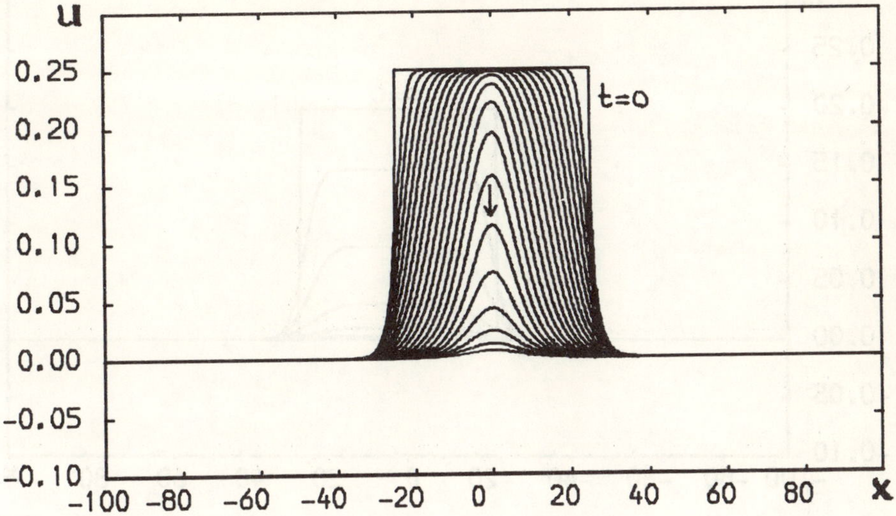

$$u_0(x) = \begin{cases} 0.25(=a) & , \quad -25 \le x \le 25 \\ 0 & , \quad \text{otherwise.} \end{cases}$$

Fig.(3) Slow extinction of the initial pulse.

(a being the threshold parameter) which is illustrated by our numerical solutions in Figs.(2-4). Recently, the following exact solutions

$$\text{(i)} \quad u = a/\{1 + a \exp(\frac{-a}{\sqrt{2}}\xi)\} \quad , \quad c = \sqrt{2}(1-{}^a/_2) \tag{9}$$

- sub-threshold front -

and

$$\text{(ii)} \quad u = \{1 + a(1-a)\exp(-\lambda\xi)\}/_{\{1 + (1-a)\exp(-\lambda\xi)\}} \tag{10}$$

$$\lambda = (1-a)/_{\sqrt{2}} \quad , \quad c = -\frac{1}{\sqrt{2}}(1+a)$$

- super-threshold front -

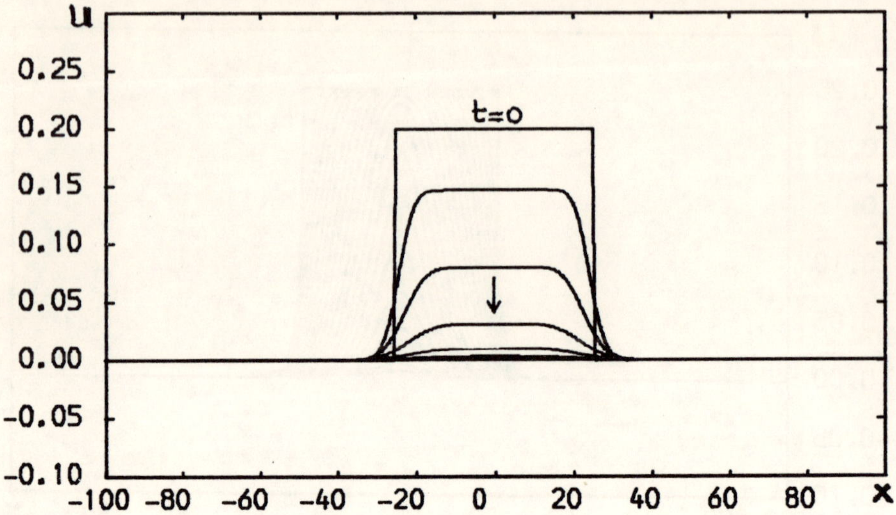

$$u_o(x) = \begin{cases} 0.2(< a=0.25) & , \quad -25 \le x \le 25 \\ 0 & , \quad \text{otherwise.} \end{cases}$$

Fig.(4) Rapid extinction of the initial pulse.

have also been obtained (Sleeman and Manoranjan [12]).

3. FITZHUGH-NAGUMO SYSTEM

The rest states of this system (2) are given by

$$u = 0 \ , \ u = \tfrac{1}{2}\{(1+a) \pm ((1-a)^2 - 4/d)^{\frac{1}{2}}\} \ .$$

If we assume that

$$d < \left\{ \frac{2}{(1-a)} \right\}^2 \tag{11}$$

then the only real rest state is $u = 0$. Under this assumption (11) we study an initial boundary value problem with the initial conditions

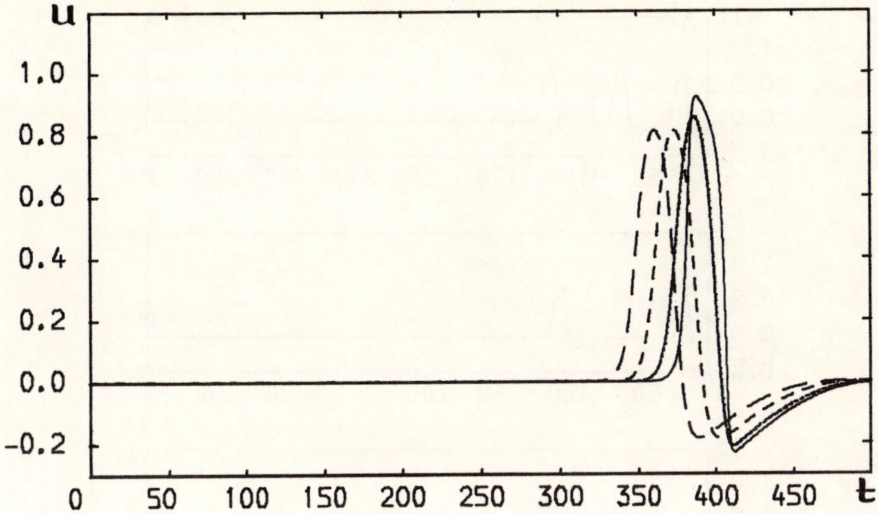

Fig.(5) Sealed boundary effect.
Curves are sketched at
x = 135, 140, 145 and 150.

$$u(x,0) = 0 = v(x,0)$$

and the boundary conditions

$$u_x(0,t) = -\tfrac{1}{2}I \quad , \quad u_x(L,t) = 0 \quad , \quad t > 0 \; , \; I > 0$$

where, L is large and I is the strength of the point stimulus at the left hand end.

If L is not large enough the right hand boundary behaves like a sealed boundary with no current leaking across the membrane. This results in an increase in peak height and a decrease in temporal displacement of the action potential u. (Fig.(5)). When $a = 0.139$, $b = 0.008$, $d = 2.54$ and $I = 0.6$, we are able to get repetitive activity and numerical experiments performed allowing only one of the above four parameters to vary in each experiment show that

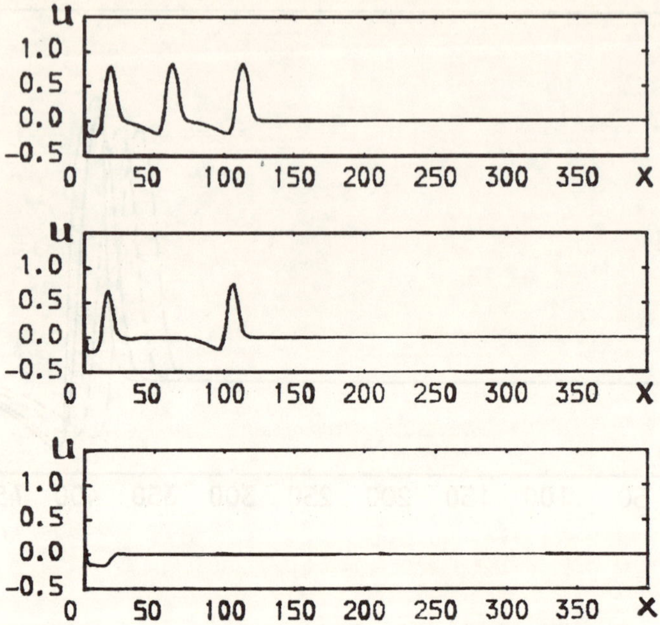

Fig.(6) Extinction of the train of pulses when a increases from 0.139.

(i) the number of pulses tends to zero with the increase in a , the threshold parameter (Fig.(6)),

(ii) the pulse solution tends to a wave front solution as $b \to 0$ (Fig.(7)),

(iii) the number of pulses tends to zero as d decreases, but an increase in d gives a front with a 'hump' (Fig.(8)), and

(iv) the repetitive activity ceases if $I > 1.2$ (nerve block); only a single pulse is obtained for these I values.

More details regarding these results can be found in Mitchell and Manoranjan [9].

Now, let us look at a linear Fitzhugh-Nagumo model written as

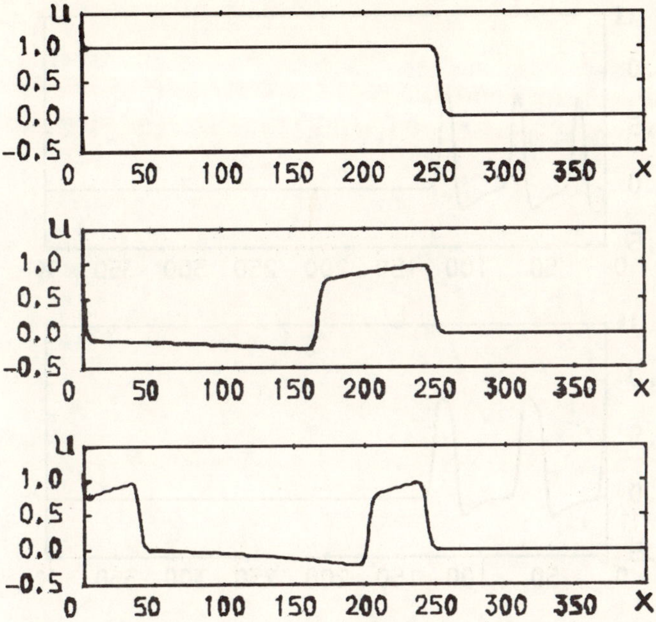

Fig.(7) Formation of repetitive activity as b increases from zero.

$$u_t = u_{xx} + f(u) - v \qquad -\infty < x < \infty$$

$$v_t = \sigma u - \gamma v \qquad 0 < a < \tfrac{1}{2} \tag{12}$$

$$t \geq 0$$

$$\sigma > 0\ ,\ \gamma \geq 0$$

$$\text{where, } f(u) = \begin{cases} -u & , \ 0 \leq u \leq a\tan\theta/(1+\tan\theta) \\ \tan\theta(u-a), & a\tan\theta/(1+\tan\theta) \leq u \leq \dfrac{(a\tan\theta+1)}{(1+\tan\theta)} \\ 1-u & , \ \dfrac{(a\tan\theta+1)}{(1+\tan\theta)} \leq u \end{cases}$$

and $\theta \in (0, \pi/2)$.

The rest states of this model are given by

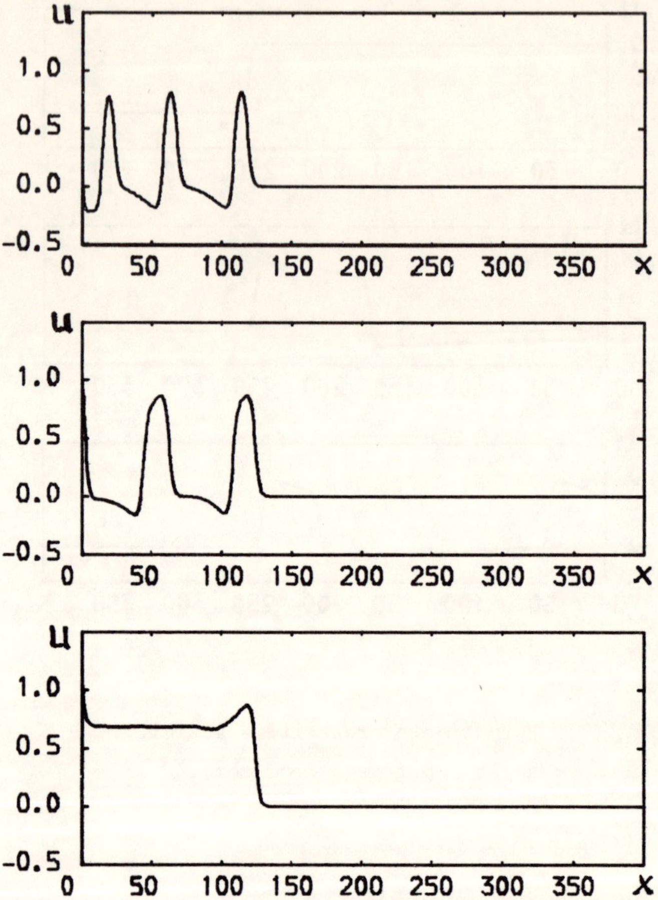

Fig.(8) Change of shape in the wave as d increases to 5.8 (which violates (11)).

$$v = f(u)$$

$$v = \frac{\sigma}{\gamma} u$$

and provided that

$$\sigma/\gamma < \frac{(1-a)\tan\theta}{(a\tan\theta+1)} < \tan\theta$$

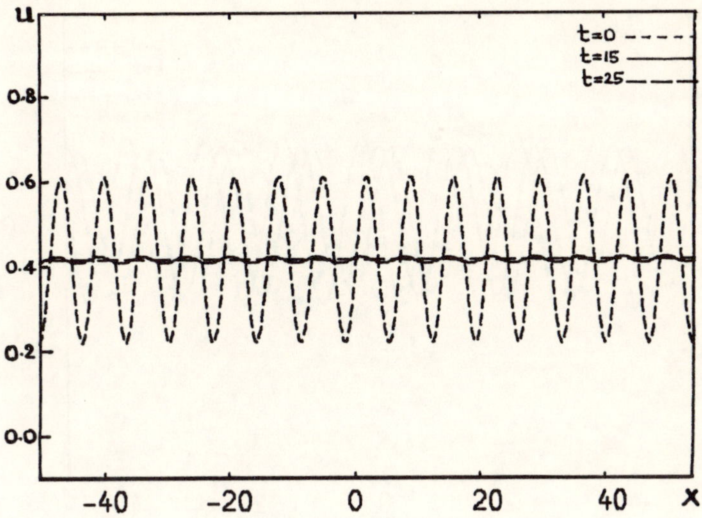

Fig.(9) Solution spirals into the intermediate rest state.
$(\tan\theta = 0.5,\ {}^{2\pi}/T_0 = 0.9 > 0.6 = (\tan\theta - \gamma)^{\frac{1}{2}})$

we have three real distinct rest states namely,

$$u = 0, \quad u = \gamma\, a \tan\theta/(\gamma\tan\theta - \sigma) \quad \text{and} \quad u = \gamma/(\gamma+\sigma)\ .$$

This system (12) was studied recently by Sleeman [11] and he showed the existence of a two parameter family of small amplitude wave solutions of period T of the form

$$u(kx + wt) = u(\xi) = \frac{\gamma a \tan\theta}{(\gamma\tan\theta - \sigma)} + A \sin \frac{2\pi}{T}(\xi + \tau) \qquad (13)$$

when $\gamma < {}^{\sigma}/\gamma < (1-a)\tan\theta/(a\tan\theta+1)$. (14)

where , $\xi = kx + wt$

$$k^2 = (\tan\theta - \gamma)T^2/4\pi^2$$

$$\omega^2 = (\sigma - \gamma^2)T^2/4\pi^2$$

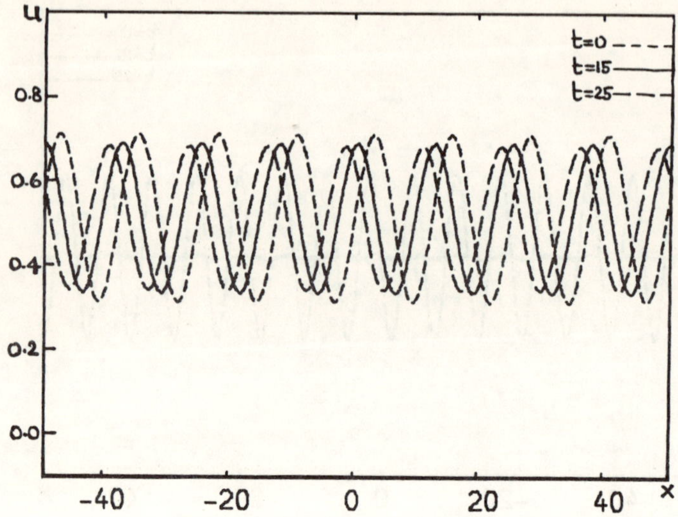

Fig.(10) Neutrally stable solution.

$(\tan\theta = 0.39, \ {}^{2\pi}/T_o = 0.5 = (\tan\theta-\gamma)^{\frac{1}{2}})$

and $A \le \min. \left[\frac{a\tan\theta+1}{1+\tan\theta} - \frac{\gamma a\tan\theta}{\gamma\tan\theta-\sigma}, \frac{\gamma a\tan\theta}{\gamma\tan\theta-\sigma} - \frac{a\tan\theta}{1+\tan\theta} \right]$. (15)

Further, looking at the stability of these periodic waves Sleeman [11] was able to prove that if $u(x,t)$ is the solution of the pure initial value problem (12) with $u(x,0) = u_o(x)$, $v(x,0) = v_o(x)$, where $u_o(x)$ is of period T_o such that

$$\frac{a\tan\theta}{1+\tan\theta} \le u_o \le \frac{a\tan\theta+1}{1+\tan\theta} \tag{16}$$

then provided that the conditions (14) and (15) hold, the behaviour of $u(x,t)$ is one of the following.

(P1) If $4\pi^2/T_o^{\ 2} > (\tan\theta-\gamma)$, $u(x,t)$ is stable and represents an oscillatory solution which spirals into the intermediate rest state $u = a\gamma\tan\theta/(\gamma\tan\theta-\sigma)$.

(P2) If $4\pi^2/T_o^{\ 2} = (\tan\theta-\gamma)$, $u(x,t)$ is neutrally stable.

(P3) If $4\pi^2/T_o^{\ 2} < (\tan\theta-\gamma)$, $u(x,t)$ is locally unstable and <u>may</u>

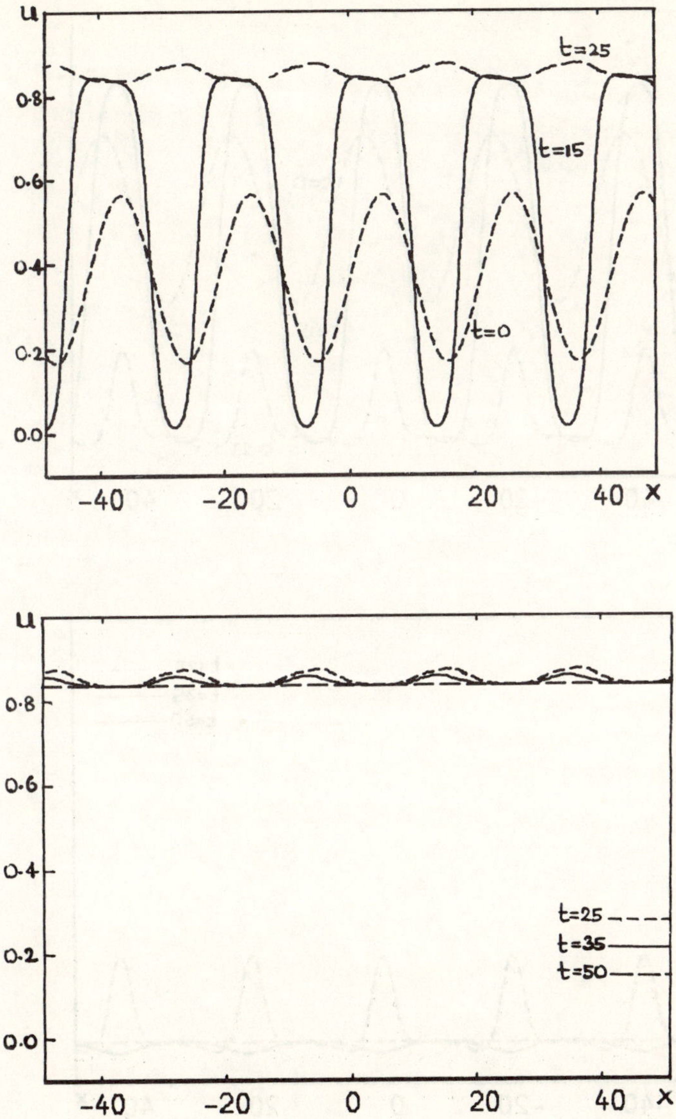

Figs.(11-12) Solution is attracted towards the upper rest state.

$(\tan\theta = 0.63,\ {}^{2\pi}/T_o = 0.3 < 0.7 = (\tan\theta - \gamma)^{\frac{1}{2}})$

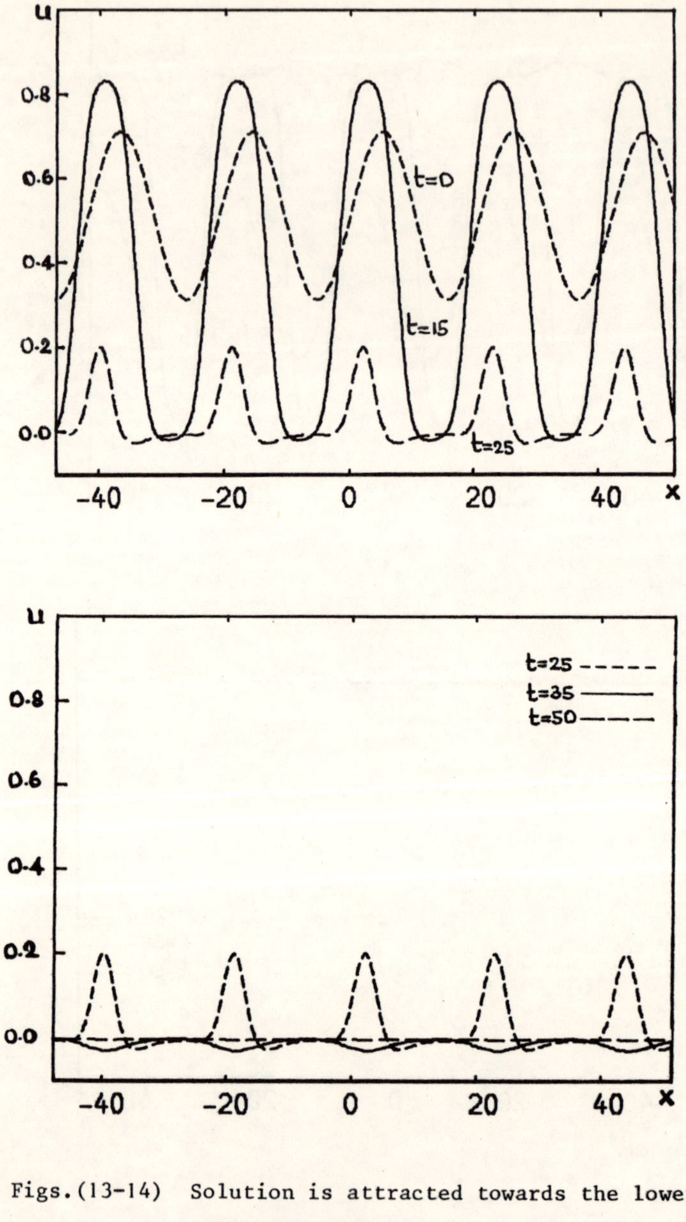

Figs.(13-14) Solution is attracted towards the lower rest state.

($\tan\theta = 0.39$, $^{2\pi}/T_o = 0.3 < 0.5 = (\tan\theta - \gamma)^{\frac{1}{2}}$)

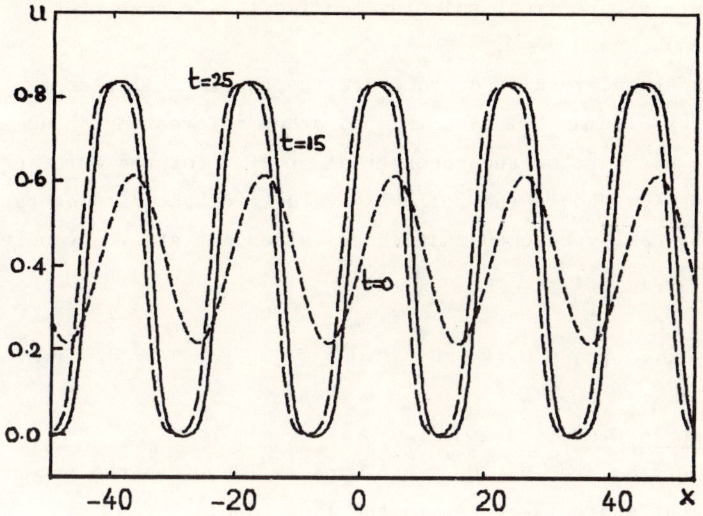

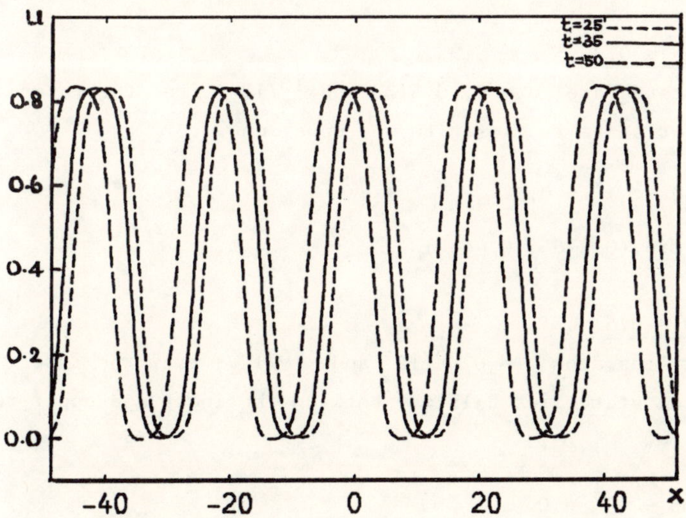

Figs.(15-16) Formation of large amplitude periodic solution.

($\tan\theta = 0.5,\ {}^{2\pi}/T_o = 0.3 < 0.6 = (\tan\theta - \gamma)^{\frac{1}{2}}$)

evolve into a wave front solution tending to the rest state $u = \gamma/(\sigma+\gamma)$, as $t \to \infty$.

Our numerical results corroborate the theoretical results obtained by Sleeman [11] and lead to other interesting phenomena which are not part of the theoretical study. For the numerical experiments we choose the following values for the parameters. (It can be easily checked that these values satisfy the conditions (14) and (15)).

$$a = 0.25 \ , \ \gamma = 0.14 \ , \ \sigma = 0.028$$

(i) <u>$\tan\theta = 0.63$</u>

The rest states are $u = 0, 0.3663$ and $10/12$, i.e. the intermediate rest state is closer to the lower one.

$$(\tan\theta - \gamma)^{\frac{1}{2}} = 0.7$$

$$A \leq \min.(0.3019, 0.3113)$$

(ii) <u>$\tan\theta = 0.39$</u>

The rest states are $u = 0, 0.5132$ and $10/12$, i.e. the intermediate rest state is closer to the upper one.

$$(\tan\theta - \gamma)^{\frac{1}{2}} = 0.5$$

$$A \leq \min.(0.2764, 0.4430)$$

(iii) <u>$\tan\theta = 0.5$</u>

The rest states are $u = 0, 5/12$ and $10/12$, i.e. the intermediate rest state lies half way between the lower and upper rest states.

$$(\tan\theta - \gamma)^{\frac{1}{2}} = 0.6$$

$$A \leq 1/3$$

We choose $A = 0.2$, which satisfies the conditions in all the three cases above.

In each case, when $\frac{2\pi}{T_0} \geq (\tan\theta-\gamma)^{\frac{1}{2}}$, the numerical results obtained confirm (P1) and (P2). Fig.(9) illustrates (P1) and

Fig.(10), (P2). However, in the case of $\frac{2\pi}{T_o} < (\tan\theta-\gamma)^{\frac{1}{2}}$, we get a different result in each case, a behaviour not pointed out by Sleeman [11].

In case (i), the solution, after leaving the range (16) is attracted towards the upper rest state and eventually takes the value of the upper rest state (Figs.(11-12)).

In case (ii), the solution is attracted towards the lower rest state and eventually takes the value of the lower rest state (Figs.(13-14)).

In case (iii), the solution grows out of range (16), but remains as a periodic solution of large amplitude oscillating between the lower and upper rest states (Figs.(15-16)). This means that when $\frac{2\pi}{T_o} < (\tan\theta-\gamma)^{\frac{1}{2}}$, the behaviour of the solution $u(x,t)$ is being dictated by the placing of the intermediate rest state.

In conclusion, it is worth commenting that we feel the above findings may be taken seriously if the biologists can find such behaviour in their experiments.

REFERENCES

1. Ablowitz, M.J., Zeppetella, A., Explicit solutions of Fisher's equation for a special wave speed, Bull. Math. Biol., 41, 835-840 (1979).

2. Aronson, D.G., Weinberger, H.G., Nonlinear diffusion in population genetics, Lecture Notes in Mathematics 446, Springer-Verlag, Berlin (1975).

3. Fife, P.C., McLeod, J.B., The approach of solutions of non-linear diffusion equations to travelling front solutions, Arch. Rat. Mech. Anal., 65, 335-361 (1977).

4. Fife, P.C., McLeod, J.B., A phase plane discussion of convergence to travelling fronts for nonlinear diffusion M.R.C. Technical Summary Report 1986 (1979).

5. Fisher, R.A., The Advance of Advantageous Genes, Ann. of Eugenics, 7, 355-369 (1937).

6. Kolmogorov, A.N., Petrovskii, I.G., Piskunov, N.S., A study of the equation of diffusion with increase in the quantity of matter and its application to biological problem, Bjul Moskouskogo Gos Univ. 1, 1-26 (1937).

7. Larson, D.A., Transient bounds and time asymptotic behaviour of solutions to non linear equations of Fisher type, SIAM J. Appl. Math., 34, 93-103 (1978).

8. McKean, H.P., Nagumo's equation, Advances in Maths. 4, 209-223 (1970).

9. Mitchell, A.R., Manoranjan, V.S., Finite element studies of reaction-diffusion, in The Mathematics of Finite Elements and Applications, MAFELAP 1981 (Ed. J.R. Whiteman), Academic Press, London (1982).

10. Rothe, F., Convergence to travelling fronts in semilinear parabolic equations, Proc. Roy. Soc. Edin., A80, 213-234, (1978).

11. Sleeman, B.D., Small amplitude periodic waves for the Fitzhugh-Nagumo equations, J. Math. Biology, 14, 309-325 (1982).

12. Sleeman, B.D., Manoranjan, V.S., The Reaction-Diffusion Equation $u_t = u_{xx} + F(u)$ (preprint) (1982).

13. Uchiyama, K., The behaviour of solutions of the equation of Kolmogorov-Petrovsky-Piskunov, Proc. Japan Acad., 53, Ser.A, 225-228 (1977).

Dr V S Manoranjan
Department of Mathematical Sciences
The University
Dundee DD1 4HN
Scotland

Dynamical equations for the stride swing in human walking

E.A. MARSHALL

1. INTRODUCTION

The study of the mechanics of animal locomotion is of interest to a number of disciplines. To the applied mathematician it constitutes a relatively unexplored and not well understood problem in the dynamics of articulated rigid bodies with constrained input controls. The medical engineer is concerned with the topic because of its relevance to the determination of limb and joint forces and motions in human gait. These are important considerations in the design and function of prostheses and in the diagnosis of, and monitoring of treatment for, orthopaedic disabilities. Recently palaeontologists have attempted to deduce, using mathematical methods, the morphology and locomotory characteristics of extinct animals known by their fossilised remains. Athletics is another activity in which locomotion is involved and, rather surprisingly perhaps, there has been collaboration between athletics coaches and mathematicians in attempts to determine the theoretical optimum performance in certain events.

Human walking is perhaps the simplest gait and an obvious topic for study. The few previous theoretical investigations that have considered this problem have employed two main approaches. One involves the use of mechanical energy principles and attempts to relate this to the metabolic energy expended in the motion. The other applies Newtonian dynamical equations to simple inertial models of jointed rigid bodies and seeks to determine the motion of the components in the action. Alexander [1] uses both approaches. His inertial model is an inverted simple pendulum in which the whole body mass is regarded as being concentrated in a single particle. Maillardet [2] investigates the in-stride motion of the swinging (ie. non weight-bearing) leg, by using the model of jointed compound pendula. Alexander's inertial model is used by Trowbridge and Paish [3] in a study of race walking mechanics. An objective of this is the determination of the theoretical maximum speed which a race walker can attain before dynamical effects cause 'lifting' of the supporting foot. When this occurs both feet are simultaneously out of contact with the ground and, according to current athletics definitions the athlete is considered to be running. Terrestrial human locomotion has also been studied by Cotton [4] and by Reynolds [5] who propose inertial models which are more comprehensive than those previously employed.

Apart from any reservations concerning the suitability of the inertial models of the anatomy employed, a number of theoretical points arise regarding the above treatments. None properly consider the three-dimensional nature of the action. In instances dynamical equations are incorrectly formulated relative to non-inertial reference axes. Some consider limbs to be freely jointed and ignore the effects that muscle control and anatomically limited joint movement have on the motion. In Maillardet's investigation this leads to unacceptably large knee hyperextension. Small stride angle approximations are also employed, either explicitly for the integration of differential equations, or less obviously elsewhere in analyses. This approximation, and the highly simplified model, account for the considerable underestimate of the maximum attainable race walking speed obtained by Trowbridge and Paish.

It is clear from the preceding discussion that the development of a general dynamical theory of gait is of interest. This has not been attempted previously. In this paper, for a comprehensive inertial model, mathematically correct three-dimensional equations of motion for the stride in human walking are derived. Functions characterising the gait style which require specification to complete the formulation are discussed as are ground contact conditions which limit the considered action.

2. CHARACTERISTICS OF HUMAN WALKING GAIT

In normal walking gait over level ground the body is carried vertically and during the stride the last contact of the free leg with the ground is through the ball of the foot with this slightly extended. The free leg enters the swing phase in this position. To accommodate the foot extension the knee flexes and straightens as the leg reaches the leading position prior to heel strike. Consequently the geometry of the legs, varies slightly throughout the stride. During walking the hips and upper body move along trajectories which are composed of a series of rising and falling arcs which have a single maximum in each stride swing. With the weight-bearing leg held perfectly stiff this occurs when the weight-bearing hip is vertically above the supporting foot [6]. The normal human gait is very close to this idealised 'stiff' walk [7]. In normal gait slight pelvic rocking may occur so that the line joining the hip locating acetabulae is inclined at a small angle to the horizontal plane.

To balance the angular momentum associated with the rotation of the legs the arms swing. This tends to be in antiphase to the stride. For fast walking, and in particular in competitive racing, maximum speed attainable without loss of ground contact is increased by the adoption of a gait in which the pelvis rotates about a vertical axis. As a result the hips are successively thrust forward in the direction of travel. This flattens the trajectories of the pelvic centre and upper body and so reduces the vertical accelerations [8]. Generation of this hip displacement requires vigorous sympathetic swing of the arms, shoulders and upper torso.

3. THE INERTIAL MODEL

In modelling the walking action it is noticed that the inertial properties of the legs, and the instride hip-ground contact distance are not substantially affected by moderate instride knee flexions and foot extensions. Also, due to their bilateral positioning and the phases of their movements in normal gait, the contribution of the angular velocities of arms, shoulders and upper-torso to the angular momentum of the system is principally to the vertical component. Consequently, and because of considerations given in Section 2, a model is adopted in which the legs are idealised as quasi-linear, non-uniform, inertial elements of length ℓ moveably jointed to the pelvis at the acetabulae A_L, A_R. The spacing between A_L,A_R is denoted by 2h. The feet are treated as blocks with curved undersurfaces. The total mass of a leg is m_ℓ and the distance of its mass centre G from the foot is $\lambda\ell$. The radius of gyration about the mass centre is $\alpha\ell$. The pelvis and lower torso are represented by a particle of mass m_p located at the mid-point P of A_LA_R whilst m_b denotes the mass of the upper torso, neck, head and arms, the mass centre B of which is located at a distance b vertically above P. The moment of inertia of these body parts about the vertical axis through B is denoted by I_b. Due to the changing configuration of the arms I_b strictly has a component which varies throughout the stride swing. To avoid excessive complication in the model, as a consequence of considerations mentioned above, the contribution to the system angular momentum from the rotations of arms, upper torso, etc is included by the addition of a term of magnitude $\dot{\gamma}I_b$ to its vertical component, where $\dot{\gamma}$ is a rotation rate characterising their motion about the vertical axis through B. γ could be interpreted as the angle, measured positive clockwise when viewed vertically downwards, which the line joining the mass centre of the arms makes with the direction of travel. The dynamical elements in this model are shown in Fig 1.

4. REFERENCE COORDINATES

In normal walking conditions both lifting and slipping of the weight bearing foot are absent. In this case the point of contact between foot and ground is a suitable choice for the origin O of a basic inertial reference frame for the specification of in-stride configuration of limb and body elements. Right-handed Cartesian axes Oxyz are chosen in which Ox lies in the direction of travel and Oy is vertically upwards. To specify the configurations, further, non-inertial, reference frames $A_Lx'y'z'$, $A_Rx''y''z''$ are introduced

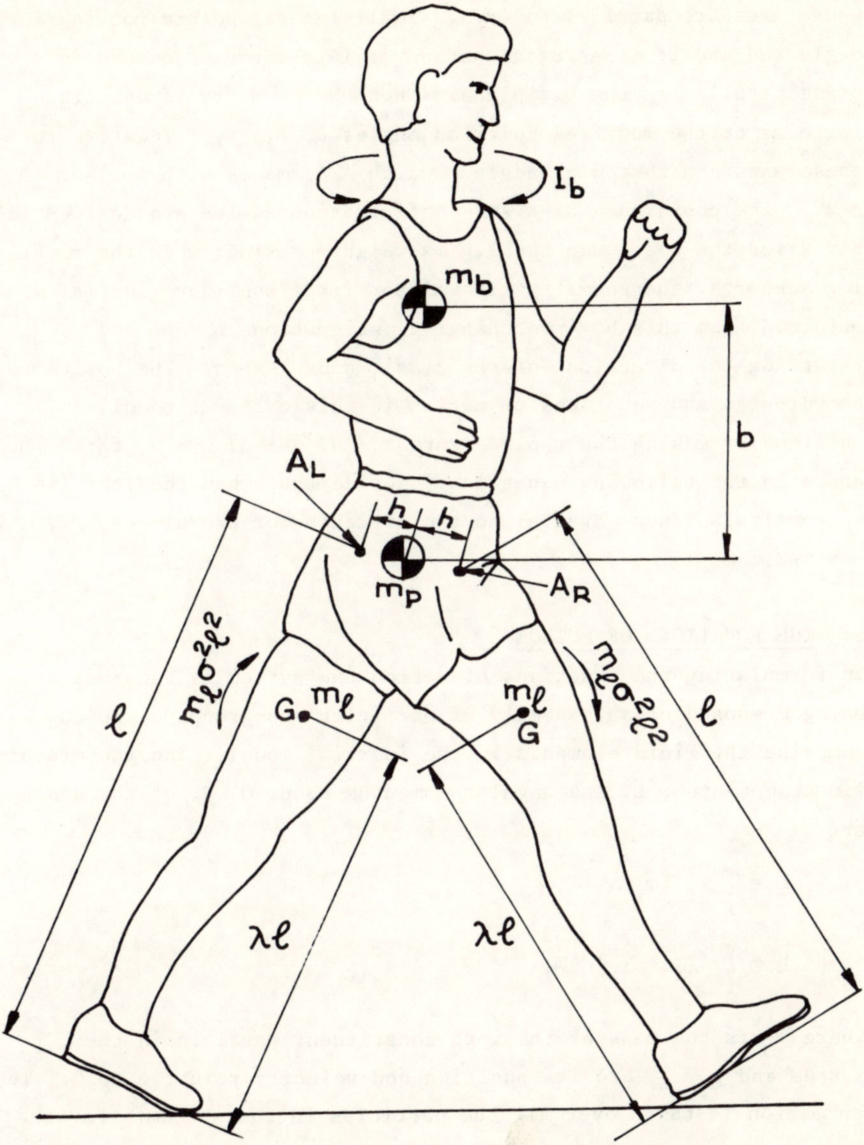

FIG 1. THE INERTIAL MODEL OF HUMAN STRIDE DYNAMICS

whose axes are parallel to Oxyz. Small lateral pelvic rocking is neglected and it is assumed that any pelvic rotation occurs in a plane parallel to the Oxz plane. Then the geometry is described in terms of the modified Eulerian angles θ, ϕ,; η, ψ relative to these axes and the polar angle α which $A_L A_R$ makes with $A_L z'$ or $A_R z''$. The coordinate axes and configuration angles are defined in Fig 2 for the case when the leg is weight-bearing; when the right leg supports, the geometry in a mirror image configuration is obtained from this by interchanging designations A_L, A_R and reversing the directions of the z, z', and z" axes. The positional coordinates and equations of motion in this case are readily obtained by making the replacements $\pi - \theta$, $\pi - \eta$, $\pi - \alpha$, for θ, η and α in the following equations. Notice that when the legs lie in vertical planes parallel to the direction of travel $\theta = \pi/2$, $\eta = \pi/2$.

5. THE EQUATIONS OF MOTION

In formulating the equations of motion the system is regarded as being composed of an assembly of particles sub-groups of which comprise the rigid elements in the inertial model. The vectors of linear momentum, $\underset{\sim}{L}$, and angular momentum about O, $\underset{\sim}{H}$, of the system are

$$\underset{\sim}{L} = \sum_i v_i \, \delta m_i \tag{1}$$

$$\underset{\sim}{H} = \sum_i (\underset{\sim}{r}_i \times \underset{\sim}{v}_i) \, \delta m_i, \tag{2}$$

where δm_i is the mass of the i-th constituent particle in the system and $\underset{\sim}{r}_i$, $\underset{\sim}{v}_i$ are its position and velocity relative to O. The summation is taken over all the particles in the system. For the mechanical system shown in Fig 1 in the configuration specified by the angular coordinates defined in Fig 2 the linear momentum calculated according to equation (1) is

$$\underset{\sim}{L} = \ell \underset{\approx}{\Omega} \underset{\sim}{K}^L, \tag{3}$$

where

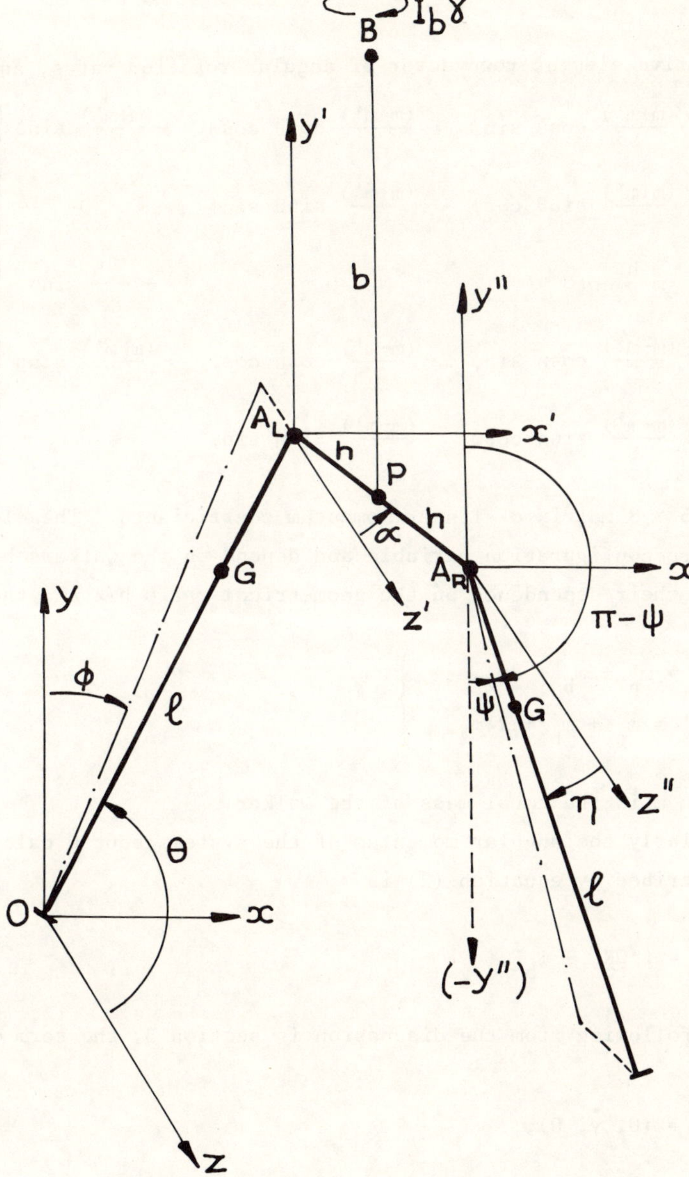

FIG 2 COORDINATE SYSTEMS AND CONFIGURATION ANGLES

$$\underset{\sim}{\Omega} = (\dot{\theta},\ \dot{\phi},\ \dot{\alpha},\ \dot{\eta},\ \dot{\psi}), \tag{4}$$

is the five element row vector of angular rotation rates, and

$$\underset{\sim}{K}^{L} = \begin{bmatrix} \frac{(m+m')}{2}\cos\theta\ \sin\phi & \frac{(m+m')}{2}\cos\theta\ \cos\phi & -\frac{(m+m')}{2}\sin\theta \\ \frac{(m+m')}{2}\sin\theta\ \cos\phi & -\frac{(m+m')}{2}\sin\theta\ \sin\phi & 0 \\ m\frac{h}{\ell}\cos\alpha & 0 & -m\frac{h}{\ell}\sin\alpha \\ \frac{(m-m')}{2}\cos\eta\ \sin\psi & -\frac{(m-m')}{2}\cos\eta\ \cos\psi & -\frac{(m-m')}{2}\sin\eta \\ \frac{(m-m')}{2}\sin\eta\ \cos\psi & \frac{(m-m')}{2}\sin\eta\ \sin\psi & 0 \end{bmatrix} \tag{5}$$

is the 5 × 3 matrix of linear momentum coefficients. The elements of $\underset{\sim}{K}^{L}$ are configuration variable and depend on the walkers build through their dependence on the geometrical ratio h/ℓ and the mass groups

$$\left.\begin{aligned} m &= m_p + m_b + 2m_\ell, \\ m' &= m_p + m_b + 2\lambda m_\ell, \end{aligned}\right\} \tag{6}$$

of which m is the total mass of the walker.

Similarly the angular momentum of the system about 0 calculated as prescribed by equation (2) is

$$\underset{\sim}{H} = \ell^2 \underset{\sim}{\Omega}\underset{\sim}{K}^{A} - I_b\dot{\underset{\sim}{\Gamma}}, \tag{7}$$

where, following from the discussion in section 3, the term $- I_b\dot{\underset{\sim}{\Gamma}}$, with

$$\dot{\Gamma} = (0,\ \dot{\gamma},\ 0), \tag{8}$$

is included to allow for the swing of the arms, shoulders etc. In equation (7) $\underset{\sim}{K}^{A}$ is a 5 × 3 matrix of angular momentum coefficients with elements

$$K^A_{11} = -m^* \cos\phi - m\frac{h}{\ell}\cos\theta\cos\phi\cos\alpha - m_b\frac{b}{\ell}\sin\theta$$

$$+\frac{(m-m')}{2}(\sin\theta\sin\eta\cos\psi - \cos\theta\cos\eta\cos\phi),$$

$$K^A_{21} = \sin\theta\sin\phi\,\{m^*\cos\theta + m\frac{h}{\ell}\sin\alpha + \frac{(m-m')}{2}\cos\eta\},$$

$$K^A_{31} = \sin\alpha\,\{-m\frac{h}{\ell}\sin\theta\cos\phi - m_b\frac{bh}{\ell^2} + (m-m')\frac{h}{\ell}\sin\eta\cos\psi\},$$

$$K^A_{41} = \hat{m}\cos\psi + \frac{(m-m')}{2}\{\cos\eta\cos\psi\,(\cos\theta + \frac{2h}{\ell}\cos\alpha) - \sin\theta\cos\phi\sin\eta\},$$

$$K^A_{51} = -\sin\eta\sin\psi\,\{\hat{m}\cos\eta + \frac{(m-m')}{2}(\cos\theta + \frac{2h}{\ell}\cos\alpha)\},$$

$$K^A_{12} = m^*\sin\phi + m\frac{h}{\ell}(\sin\theta\sin\alpha + \cos\theta\sin\phi\cos\alpha)$$

$$+\frac{(m-m')}{2}(\cos\eta\cos\theta\sin\phi + \sin\theta\sin\eta\sin\psi),$$

$$K^A_{22} = \sin\theta\cos\phi\,\{m^*\cos\theta + m\frac{h}{\ell}\cos\alpha + \frac{(m-m')}{2}\cos\eta\},$$

$$K^A_{32} = (m+2m_\ell)\frac{h^2}{\ell^2} + m\frac{h}{\ell}(\sin\theta\sin\phi\sin\alpha + \cos\theta\cos\alpha)$$

$$+(m-m')\frac{h}{\ell}(\cos\alpha\cos\eta + \sin\alpha\sin\eta\sin\psi),$$

$$K^A_{42} = \hat{m}\sin\psi + \frac{(m-m')}{2}\{\cos\eta\sin\psi\,(\cos\theta + \frac{2h}{\ell}\cos\alpha)$$

$$+\sin\eta\,(\sin\theta\sin\phi + \frac{2h}{\ell}\sin\alpha)\},$$

$$K^A_{52} = \sin\eta\cos\psi\,\{\hat{m}\cos\eta + \frac{(m-m')}{2}(\cos\theta + \frac{2h}{\ell}\cos\alpha)\},$$

$$K^A_{13} = \cos\theta\,\{m\frac{h}{\ell}\cos\phi\sin\alpha - m_b\frac{b}{\ell}\sin\phi + \frac{(m-m')}{2}\sin\eta\sin\overline{\phi+\psi}\},$$

$$K^A_{23} = \sin\theta\,\{-m^*\sin\theta - m\frac{h}{\ell}\sin\phi\sin\alpha - m_b\frac{b}{\ell}\cos\phi$$

$$+\frac{(m-m')}{2}\sin\eta\cos\overline{\phi+\psi}\},$$

$$K^A_{33} = \cos\alpha\,\{-m\frac{h}{\ell}\sin\theta\cos\phi - m_b\frac{bh}{\ell^2} + (m-m')\frac{h}{\ell}\sin\eta\cos\psi\},$$

$$K^A_{43} = \frac{(m-m')}{2}\cos\eta\,\{\sin\theta\sin\overline{\phi+\psi} - \frac{2h}{\ell}\cos\psi\sin\alpha\},$$

$$K^A_{53} = \sin\eta \; \{\hat{m} \sin\eta - \frac{(m-m')}{2} \; (\sin\theta \cos\overline{\phi+\psi} - \frac{2h}{\ell} \sin\psi \sin\alpha) \; ,$$

equs. (9) 1-15

which, like $\underset{\sim}{K}^L$, is also configuration variable and depends on the walkers build because of its dependence on m, m', $\frac{h}{\ell}$ and also because it involves the additional geometrical ratio $\frac{b}{\ell}$ and the mass groups

$$\left.\begin{aligned} m^* &= m_p + m_b + (1 + \lambda^2 + \sigma^2)m_\ell , \\ \hat{m} &= \{(1 - \lambda)^2 + \sigma^2\}m_\ell , \quad (= m^* - m'). \end{aligned}\right\} \tag{10}$$

The total external force

$$\underset{\sim}{F} = \sum_i \underset{\sim}{F}_i \tag{11}$$

and resultant moment about O

$$\underset{\sim}{G} = \sum_i \underset{\sim}{r}_i \times \underset{\sim}{F}_i \tag{12}$$

of the external (gravity) forces are

$$\underset{\sim}{F} = (R_1, R_2 - mg, R_3), \tag{13}$$

$$\underset{\sim}{G} = (G_1, 0, G_3), \tag{14}$$

where R_1, R_2, R_3 are the components in the x, y, z directions of the (unknown) reaction at O on the supporting foot, and

$$\left.\begin{aligned} G_1 &= g\ell \left\{ \frac{(m+m')}{2} \cos\theta + m \frac{h}{\ell} \cos\alpha + \frac{(m-m')}{2} \cos\eta \right\} , \\ G_3 &= - g\ell \left\{ \frac{(m+m')}{2} \sin\theta \sin\phi + m \frac{h}{\ell} \sin\alpha + \frac{(m-m')}{2} \sin\eta \sin\psi \right\} . \end{aligned}\right\} \tag{15}$$

Notice that in the absence of ground friction couples the component of G along the Oy axis is zero.

The general equations of motion of the system are obtained by using the expressions for $\underset{\sim}{L}$, $\underset{\sim}{H}$, $\underset{\sim}{F}$, $\underset{\sim}{G}$ derived above in the fundamental dynamical equations

$$\underset{\sim}{F} = \dot{\underset{\sim}{L}}, \quad \underset{\sim}{G} = \dot{\underset{\sim}{H}}. \tag{16}$$

The resulting equations are valid for actions where there is no linear motion of the supporting foot relative to the ground.

For this condition to prevail R_1, R_2, R_3 must satisfy the conditions

$$\left.\begin{array}{l} R_2 > 0, \\ R_1^2 + R_3^2 < F_a^2, \end{array}\right\} \tag{17}$$

where F_a is the maximum allowable horizontal component of the contact reaction. For hard dry surfaces

$$F_a = \mu R_2 \tag{18}$$

where μ is the coefficient of friction between foot and ground. On soft, muddy terrain F_a is determined by the foot contact area and the rheological properties of the ground surface. In this case F_a can be increased by ridge-soled or studded footwear. Consequently except for walking on hard polished surfaces it is expected that the limiting form

$$R_2 = 0 \tag{19}$$

of relation (17), will be the mechanical condition which determines the maximum walking speed.

Because of the unwieldy form of the matrices $\underset{\sim}{K}^L$, $\underset{\sim}{K}^A$, particularly $\underset{\sim}{K}^A$, the general equations of motion are somewhat unmanageable. However, it is clear that in conventional walking styles the lateral deviations

$$\varepsilon = \theta - \frac{\pi}{2}\,, \quad \delta = \eta - \frac{\pi}{2}\,, \tag{20}$$

of the legs from the vertical are sufficiently small for the approximations

$$\left.\begin{array}{ll} \sin\varepsilon \simeq \varepsilon\,, & \cos\varepsilon \simeq 1, \\ \sin\delta \simeq \delta\,, & \cos\delta \simeq 1, \end{array}\right\} \tag{21}$$

to describe satisfactorily the configuration. The equations of motion for such gaits are presented here. Notice that no restrictions are introduced regarding the magnitudes of the forward stride angles ϕ, ψ or the pelvic rotation α.

By using approximations (21) and neglecting squares and products, and higher order terms of ε, δ, $\dot{\varepsilon}$, $\dot{\delta}$ it is found that for this case equations (3), (4), (5), (7) and (9) for the linear and angular momentum reduce to

$$\underset{\sim}{L} = \ell\underset{\sim}{\omega}\underset{\sim}{k}^{L}, \tag{22}$$

$$\underset{\sim}{H} = \ell^2\underset{\sim}{\omega}\underset{\sim}{k}^{A} - I_b\dot{\underset{\sim}{\Gamma}}, \tag{23}$$

where

$$\underset{\sim}{\omega} = (\dot{\varepsilon}, \dot{\phi}, \dot{\alpha}, \dot{\delta}, \dot{\psi}), \tag{24}$$

$$\underset{\sim}{k}^{L} = \begin{bmatrix} 0 & 0 & -\frac{(m+m')}{2} \\ \frac{(m+m')}{2}\cos\phi & -\frac{(m+m')}{2}\sin\phi & 0 \\ m\frac{h}{\ell}\cos\alpha & 0 & -m\frac{h}{\ell}\sin\alpha \\ 0 & 0 & -\frac{(m-m')}{2} \\ \frac{(m-m')}{2}\cos\psi & \frac{(m-m')}{2}\sin\psi & 0 \end{bmatrix}, \tag{25}$$

and $\underset{\sim}{k}^{A}$ is the 5×3 matrix with elements

$$k^A_{11} = - m^* \cos\phi - m_b \frac{b}{\ell} + \frac{(m-m')}{2} \cos\psi,$$

$$k^A_{21} = - \sin\phi \{m^*\varepsilon - m \frac{h}{\ell} \cos\alpha + \frac{(m-m')}{2} \delta\},$$

$$k^A_{31} = \sin\alpha \{- m \frac{h}{\ell} \cos\phi - m_b \frac{bh}{\ell^2} + (m-m') \frac{h}{\ell} \cos\psi\},$$

$$k^A_{41} = \hat{m} \cos\phi - \frac{(m-m')}{2} \cos\phi,$$

$$k^A_{51} = \sin\psi \{\hat{m}\delta + \frac{(m-m')}{2} (\varepsilon - \frac{2h}{\ell} \cos\alpha)\},$$

$$k^A_{12} = m^* \sin\phi + m \frac{h}{\ell} \sin\alpha + \frac{(m-m')}{2} \sin\psi,$$

$$k^A_{22} = - \cos\phi \{m^*\varepsilon - m \frac{h}{\ell} \cos\alpha + \frac{(m-m')}{2} \delta\},$$

$$k^A_{32} = \frac{h}{\ell} \{m(\sin\phi \sin\alpha - \varepsilon \cos\alpha) + (m-m')(\sin\psi \sin\alpha - \delta \cos\alpha) + \frac{h}{\ell} (m+2m_\ell)\},$$

$$k^A_{42} = \hat{m} \sin\psi + \frac{(m-m')}{2} (\sin\phi + \frac{2h}{\ell} \sin\alpha),$$

$$k^A_{52} = - \cos\psi \{\hat{m}\delta + \frac{(m-m')}{2} (\varepsilon - \frac{2h}{\ell} \cos\alpha)\},$$

$$k^A_{13} = 0,$$

$$k^A_{23} = - m^* - m \frac{h}{\ell} \sin\phi \sin\alpha - m_b \frac{b}{h} \cos\phi + \frac{(m-m')}{2} \cos\overline{\phi+\psi},$$

$$k^A_{33} = \frac{h}{\ell} \cos\alpha \{- m \cos\phi - m_b \frac{b}{\ell} + (m-m') \cos\psi\},$$

$$k^A_{43} = 0,$$

$$k^A_{53} = \hat{m} - \frac{(m-m')}{2} (\cos\overline{\phi+\psi} - \frac{2h}{\ell} \sin\psi \sin\alpha). \qquad \text{equ} \quad (26)_{1-15}$$

Similarly the non-zero components of the moment reduce to

$$\left.\begin{aligned} G_1 &= - g\ell \left\{ \frac{(m+m')}{2} \varepsilon - m \frac{h}{\ell} \cos\alpha + \frac{(m-m')}{2} \delta \right\}, \\ G_3 &= - g\ell \left\{ \frac{(m+m')}{2} \sin\phi + m \frac{h}{\ell} \sin\alpha + \frac{(m-m')}{2} \sin\psi \right\}. \end{aligned}\right\} \quad (27)$$

In the derivation of the equations of motion the use of subscript notation and summation convention is convenient and in this connection the row vector

$$\underset{\sim}{\xi} = (\xi_i) = (\varepsilon, \phi, \alpha, \delta, \psi), \tag{28}$$

of configuration angles is introduced. Then the scalar equations of motion are

$$\left.\begin{aligned} F_j &= \ell(\dot{\omega}_i k^L_{ij} + \omega_i \omega_p D^L_{ijp}), \\ G_j &= \ell^2(\dot{\omega}_i k^A_{ij} + \omega_i \omega_p D^A_{ijp}) - \overbrace{(I_b \dot{\Gamma}_j)}^{\bullet} \end{aligned}\right\} \tag{29}$$

where

$$\left.\begin{aligned} D^L_{ijp} &= \tfrac{1}{2}\left(\frac{\partial k^L_{ij}}{\partial \xi_p} + \frac{\partial k^L_{pj}}{\partial \xi_i}\right), \\ D^A_{ijp} &= \tfrac{1}{2}\left(\frac{\partial k^A_{ij}}{\partial \xi_p} + \frac{\partial k^A_{pj}}{\partial \xi_i}\right). \end{aligned}\right\} \tag{30}$$

Because of the form of the matrices $\underset{\sim}{k}^L$ and $\underset{\sim}{k}^A$, D^L_{ijp} and D^A_{ijp} have the properties

$$\left.\begin{aligned} D^L_{ijp} &= 0 \quad i \neq p, \\ D^A_{ijp} &= 0 \quad i \neq p, \end{aligned}\right\} \tag{31}$$

and consequently equations (29) can be written as

$$\left.\begin{aligned} F_j &= \ell(\dot{\omega}_i k^L_{ij} + \omega_i^2 E^L_{ij}) \\ G_j &= \ell^2(\dot{\omega}_i k^A_{ij} + \omega_i^2 E^A_{ij}) - \overbrace{(I_b \dot{\Gamma}_j)}^{\bullet}, \end{aligned}\right\} \tag{32}$$

where

$$\left.\begin{aligned} \omega_i^2 &= \omega_i \omega_i, \\ E_{ij}^L &= \frac{\partial k_{ij}^L}{\partial \xi_i}, \\ E_{ij}^A &= \frac{\partial k_{ij}^A}{\partial \xi_i}. \end{aligned}\right\} \quad \text{(i not summed)} \tag{33}$$

In matrix notation the equations of motion are

$$\left.\begin{aligned} \underset{\sim}{F} &= \ell(\underset{\sim}{\dot{\omega}}\underset{\sim}{k}^L + \underset{\sim}{\omega}^2 \underset{\sim}{E}^L) \\ \underset{\sim}{G} &= \ell^2(\underset{\sim}{\dot{\omega}}\underset{\sim}{k}^A + \underset{\sim}{\omega}^2 \underset{\sim}{E}^A) - \overbrace{(I_b \underset{\sim}{\dot{\Gamma}})}^{\cdot} \end{aligned}\right\} \tag{34}$$

where

$$\left.\begin{aligned} \underset{\sim}{\dot{\omega}} &= (\ddot{\varepsilon}, \ddot{\phi}, \ddot{\alpha}, \ddot{\delta}, \ddot{\psi}), \\ \underset{\sim}{\omega}^2 &= (\dot{\varepsilon}^2, \dot{\phi}^2, \dot{\alpha}^2, \dot{\delta}^2, \dot{\psi}^2), \end{aligned}\right\} \tag{35}$$

and

$$\underset{\sim}{E}^L = (E_{ij}^L) = \begin{bmatrix} 0 & 0 & 0 \\ -\frac{(m+m')}{2}\sin\phi & -\frac{(m+m')}{2}\cos\phi & 0 \\ -m\frac{h}{\ell}\sin\alpha & 0 & -m\frac{h}{\ell}\cos\alpha \\ 0 & 0 & 0 \\ -\frac{(m-m')}{2}\sin\psi & -\frac{(m-m')}{2}\cos\psi & 0 \end{bmatrix}, \tag{36}$$

$$\underset{\sim}{E}^A = (E_{ij}^A), \tag{37}$$

the elements of $\underset{\sim}{E}^A$ being

$$E_{11}^A = 0,$$

$$E_{21}^A = -\cos\phi\,\{m^*\varepsilon - m\frac{h}{\ell}\cos\alpha + \frac{(m-m')}{2}\,\delta\},$$

$$E_{31}^A = \cos\alpha\,\{-m\frac{h}{\ell}\cos\phi - m_b\frac{bh}{\ell^2} + (m-m')\frac{h}{\ell}\cos\psi\},$$

$$E^A_{41} = 0,$$

$$E^A_{51} = \cos\psi \ \{\hat{m}\delta + \frac{(m-m')}{2} \ (\varepsilon - \frac{2h}{\ell} \cos\alpha)\},$$

$$E^A_{12} = 0,$$

$$E^A_{22} = \sin\phi \ \{m^*\varepsilon - m \frac{h}{\ell} \cos\alpha + \frac{(m-m')}{2} \ \delta\},$$

$$E^A_{32} = \frac{h}{\ell} \ \{m(\sin\phi \ \cos\alpha + \varepsilon \ \sin\alpha) + (m-m')(\sin\psi \ \cos\alpha + \delta \ \sin\alpha)\},$$

$$E^A_{42} = 0,$$

$$E^A_{52} = \sin\psi \ \{\hat{m}\delta + \frac{(m-m')}{2} \ (\varepsilon - \frac{2h}{\ell} \cos\alpha)\},$$

$$E^A_{13} = 0,$$

$$E^A_{23} = - m \frac{h}{\ell} \cos\phi \ \sin\alpha + m_b \frac{b}{\ell} \sin\phi - \frac{(m-m')}{2} \ \sin\overline{\phi+\psi},$$

$$E^A_{33} = \frac{h}{\ell} \sin\alpha \ \{m \cos\phi + m_b \frac{b}{\ell} - (m-m') \cos\psi\},$$

$$E^A_{43} = 0,$$

$$E^A_{53} = \frac{(m-m')}{2} \ \{\sin\overline{\phi+\psi} + \frac{2h}{\ell} \cos\psi \ \sin\alpha\},$$

equs (38)
1-15

This completes the derivation of the equations of motion. In the absence of ground friction couples the explicit evaluation of the term

$$\overset{\bullet}{\overbrace{(I_b \dot{\underset{\sim}{\Gamma}})}} \equiv \frac{d}{dt} (I_b \dot{\underset{\sim}{\Gamma}})$$

in equation (34) with variable I_b is not required since $\dot{\Gamma}_2 = \dot{\gamma}$ is the only non-zero component of $\dot{\underset{\sim}{\Gamma}}$ and in this case the equation of angular motion about the Oy axis has the exact first integral

$$H_2 = C. \tag{39}$$

The constant C is determined from initial conditions prescribed for the action. This result applies irrespective of whether the small lateral deviation approximations (21) are employed. The presence of $\dot{\varepsilon}^2$, $\dot{\delta}^2$ in equation $(35)_2$ is not inconsistent with the approximation, used in the derivation of equations (19) - (24), that the squares and products of the small quantities ε, δ, $\dot{\varepsilon}$, $\dot{\delta}$ are negligible. This is because in the matrix products $\underset{\sim}{\omega}^2\underset{\sim}{E}^L$, $\underset{\sim}{\omega}^2\underset{\sim}{E}^A$ terms involving $\dot{\varepsilon}^2$, $\dot{\delta}^2$ only occur with zero coefficients.

The equations of motion form a system of six coupled ordinary differential equations which are linear in the angular accelerations, quadratic in the rotation rates and have configuration variable coefficients. For a given configuration solution the three equations of rectilinear motion determine the components R_1, R_2, R_3 of ground reaction required to support the stride action. With regard to the configuration solution, the five configuration angles ε, ϕ, α, δ, ψ and the arm/shoulder swing-rate $\dot{\gamma}$ are involved in the three equations of angular motion. Consequently dynamical considerations alone are insufficient for their determination. In this respect gait motion differs from classical dynamics problems involving inanimate systems, for example elastically coupled oscillators; multiple pendula. In these the interaction is purely mechanical and by using well-known techniques (eg. Lagrange's equations) as many equations of motion may be formulated as there are configuration coordinates. In walking, as in other modes of animal locomotion, the individuals motor nervous system initiates muscular contractions which do work on the mechanical components of the anatomy. In this way a desired gait and stride length are adopted. Of course these must be compatible with restrictions on joint mobility. Mechanical effects determine the action to the extent that it must satisfy the equations of motion and comply with constraints on the ground contact reactions.

6. RELATIONS CHARACTERISING THE GAIT STYLE

In inertial models for gait involving n configuration coordinates the introduction of n-3 relations characterising gait style is required. One approach is to decompose the model into its constituents and to specify n-3 variable interaction couples exerted at the joints by the muscles. These could be determined from experimentally measured muscle tensions and knowledge of the joint geometry. Then the simultaneous solution of the coupled sets of equations of motion of the individual constituents would be required. This procedure appears complicated.

A simpler but meaningful way of describing gait style, applicable to the present study, is to specify n-3 free coordinates as functions of the 3 remaining coordinates which are dynamically determined. In the present model n = 6 and so 3 such gait style functions are needed. There are 20 mathematically possible ways in which the triplet of free variables may be chosen. Because the walker appears to have full control over the style of free leg movement and because pelvic rotation appears to be voluntary, requiring vigorous arm-swing for its generation, the choice of (δ, ψ, α) as free variables appears appropriate. Then (δ, ψ, α) can be taken to be related to (ε, ϕ, γ) by relations of the form

$$\left.\begin{aligned} \delta &= \mathcal{F}_{\delta}(\varepsilon), \\ \psi &= \mathcal{F}_{\psi}(\phi), \\ \alpha &= \mathcal{F}_{\alpha}(\gamma), \end{aligned}\right\} \qquad (40)$$

where $\mathcal{F}_{\delta}$, $\mathcal{F}_{\psi}$, $\mathcal{F}_{\alpha}$ are prescribed functions which characterise the particular gait style. The representation is not unique because ε, ϕ, γ are all related in time from the dynamical solution. With the representation in equations (40) the specification of δ, $\dot{\delta}$; ψ, $\dot{\psi}$; α, $\dot{\alpha}$ at the beginning of the stride determines through equations (40), and their total time derivatives, associated values of ε, $\dot{\varepsilon}$; ϕ, $\dot{\phi}$; γ, $\dot{\gamma}$. These provide sufficient initial conditions for the solution of the well-posed system of differential equations resulting from the combination of equations $(34)_2$, (40). There does not appear to be any published data giving particular forms for the gait characterising functions for specific walking styles. Their experimental determination is non-trivial and is likely to require rolling walkway and/or cinematographic procedures.

In conventional walking the swing of arms, shoulders and upper body are of the same frequency as the stride-rate and the pelvic rotation rate (if any), but of opposite rotational sense with respect to the vertical axis. Consequently for some gait styles

$$\mathcal{F}_\alpha \equiv c_\gamma \gamma + d, \tag{41}$$

where c_γ, d are constants might give satisfactory representation of the variation of the free coordinate α. For the case where pelvic rotation is absent ($\alpha \equiv 0$) $c_\gamma = 0$ and $d = 0$ and γ is dynamically determined. Biomechanical laboratory walking trials indicate that, except in bizarre gaits with large free-leg abduction, δ is small throughout the stride-swing and deviates very little from the end of stride value. This suggests that a satisfactory approximation is

$$\mathcal{F}_\delta \equiv c_\delta \tag{42}$$

where $c_\delta < h/\ell$ is constant. With regard to $\mathcal{F}_\psi$, observation of normal human walking shows that the weight-bearing leg initially approaches the vertical position rather more rapidly than the free-leg; this then catches up and both legs pass through the vertical at about the same instant. Subsequently the free leg advances faster and the angle it swings through may be sufficiently large to require reverse swing prior to heel strike. This feature is prominent in some military marching styles, eg. goosestepping. In the end of stride double contact position with the body symmetrical with respect to the central vertical plane in the direction of travel the end of stride values ϕ_s, ψ_s of ϕ, ψ satisfy

$$\psi_s = \phi_s \tag{43}$$

Fig 3 shows schematic graphs of functions $\mathcal{F}_\psi(\phi)$ for some walking styles, of given stride length, with the above characteristics. It is clear that for moderate gait styles the main features of $\mathcal{F}_\psi$ are reasonably well described by

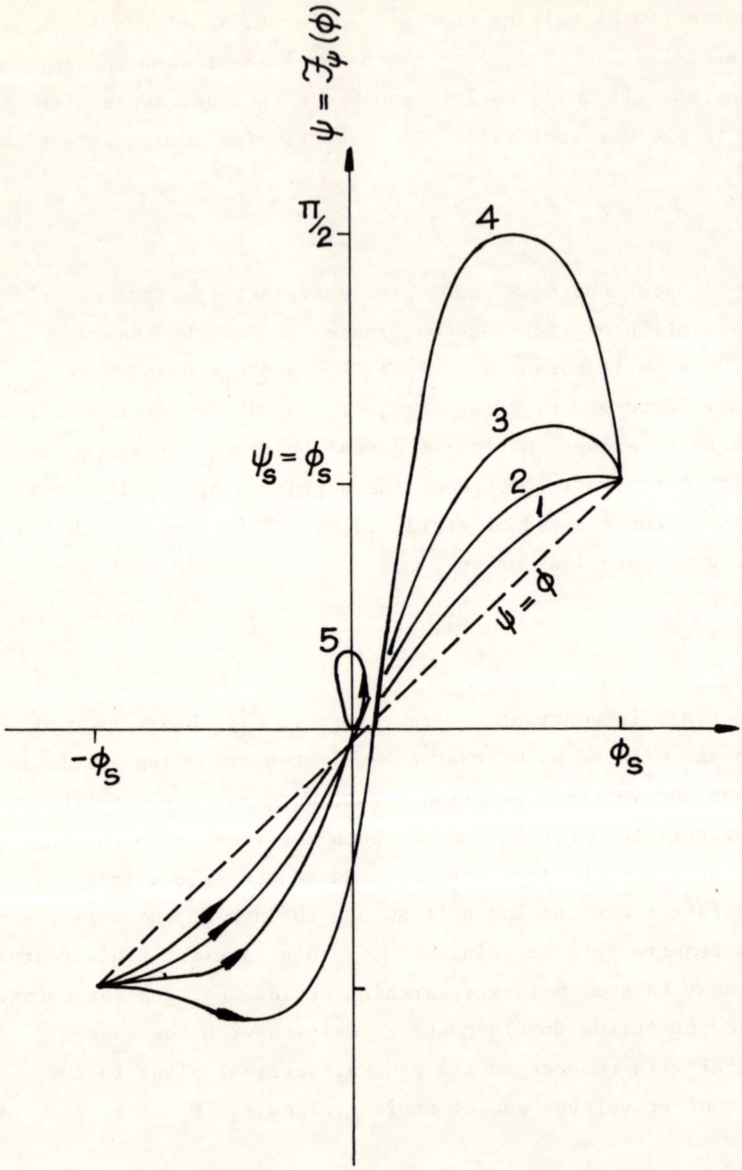

FIG 3. SCHEMATIC GRAPHS OF THE GAIT FUNCTION $\mathcal{F}_\psi(\phi)$ FOR VARIOUS GAIT STYLES

(1) Stroll (2) Normal Gait (3) Vigorous stride

(4) Goose step (5) Marking time (stiff leg)

$$\mathcal{F}_{\psi}(\phi) \equiv \phi + c_{\psi} \sin \pi\phi/\phi_s, \tag{44}$$

where c_{ψ} is constant. In relaxed walking, where free leg fore-lift is small, $c_{\psi} << 1$ and

$$\mathcal{F}_{\psi}(\phi) \simeq \phi. \tag{45}$$

The dynamical equations (32), and gait style functions equations (40) are readily transformed into a system of first-order non-linear differential equations which, together with specified initial conditions as indicated above, is suitable for numerical solution by the Runge-Kutta technique. This is currently under consideration.

REFERENCES

1. Alexander, R. McN. (1977) Mechanics and Scaling of Terrestrial Locomotion. In: Scale Effects in Animal Locomotion, ed. T.J. Pedley pp. 93-110, Academic Press, London.
2. Maillardet, F. J. (1977), The Swing Phase of Locomotion, Engineering in Medicine, 6, 67-75.
3. Trowbridge, E.A. & Paish, W. (1981) Mechanics of Athletics, Bull. Inst. Maths. Applics. 17, 144-146.
4. Cotton, A.J.W. (1980) A study of Human Terrestrial Locomotion. University of Sheffield, Dissertation.
5. Reynolds, G.D. (1981) Mathematical Models of Human Terrestrial Locomotion. University of Sheffield, Dissertation.
6. Alexander, R. McN. (1980) The Mechanics of Walking. In: Aspects of Animal Movement. eds. H.Y. Elder & E.R. Trueman. Cambridge Un. Press.
7. Carlsöö, S. (1972) How Man Moves. Heineman, London.
8. Dyson, G. (1973) The Mechanics of Athletics. London Un. Press..

E A Marshall
Department of Applied & Computational Mathematics
University of Sheffield
Sheffield S10 2TN, England.

Tractable population models with age structure

R.M. NISBET and W.S.C. GURNEY

1. INTRODUCTION: AIMS OF THEORETICAL POPULATION DYNAMICS

This being the opening talk on "Population Dynamics" and the intended audience having rather diverse interests, I propose to start with a few remarks on the objectives of population modelling and then focus on a single problem.

Irrespective of the complexity of the subsequent analysis, all population dynamics must start with a balance equation reflecting the fact that the size of a population can only change because of births, deaths, immigrations or emigrations, the vast variety of types of dynamical equations found in the literature reflecting different assumptions about these four basic processes. Population models are conveniently subdivided into two categories depending on the nature of these assumptions. "Strategic models" incorporate as little detail as possible and have the aim of generating hypotheses rather than producing testable predictions; they model "nothing in particular and do it very well" (W.S. Gilbert: Iolanthe). By contrast "tactical models" (or "simulations") exploit precise and detailed information on a particular species in order to make short term population forecasts (e.g. by actuaries on

Homo sapiens). Occasionally such models also yield scientific insight, but many deserve to be studied on Bob May's now notorious computer equipped with an on-line incinerator.[1]

Faced with models which are either untestable or unusable, the majority of ecologists dismiss mathematical models as irrelevant to their discipline. A significant minority however urge modellers to refine their art by studying *controlled laboratory populations*. We believe that this advice is sound, since it is certainly reasonable to demand conventional, falsifiable models of such populations, yet to deliver is remarkably difficult. Furthermore, by way of a glimmer on the horizon, improved models of controlled populations may help us understand better the relationship between laboratory and field - arguably one of the "ultimate questions" underlying experimental ecology.

Insect populations are a particularly attractive testing ground for our models and in this talk I shall refer on several occasions to laboratory populations which exhibit large quasi-periodic fluctuations (e.g. Fig. 1), primarily because predicting the period and amplitude of such fluctutations might reasonably be regarded as a mimimal objective. It is reassuring to find that these quantities frequently turn out to be very sensitive to assumptions on the nature of regulating mechanisms[18]; consequently modelling such oscillations should be ecologically useful and not just part of an expensive, if entertaining, game for physicists.

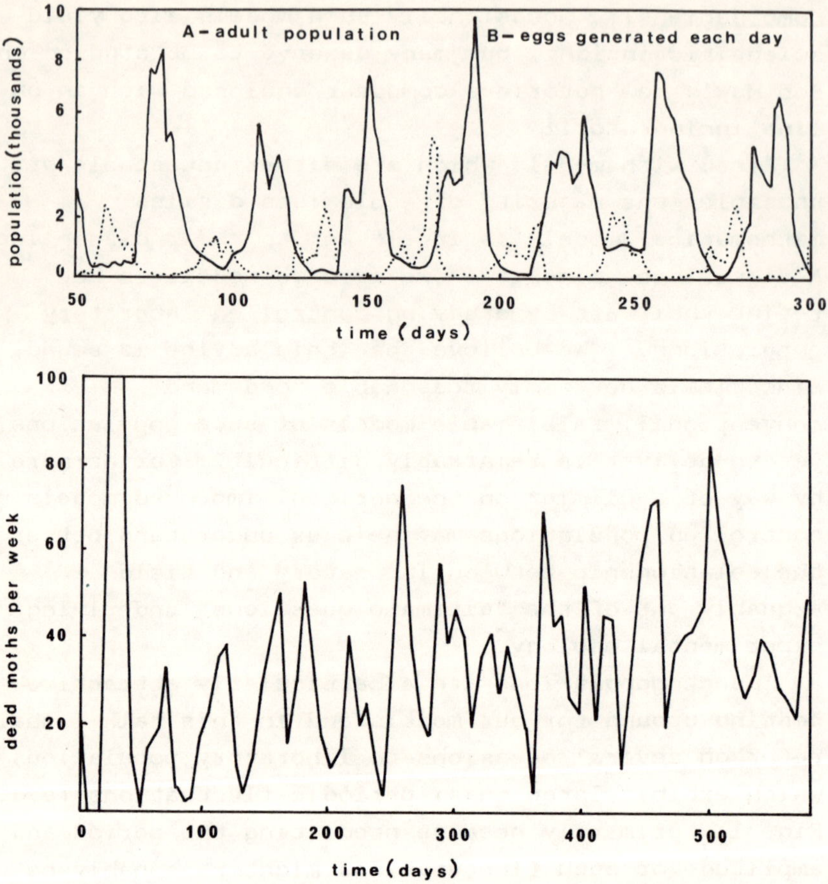

Fig.1: Large quasi-cyclic fluctuations in laboratory insect populations.
TOP - Nicholson's blowflies[2] regulated by adult competition
BOTTOM - Lawton's meal moths[3] regulated by larval competition

2. THE FORMULATION OF AGE STRUCTURE MODELS

There are two popular routes to defensible population models incorporating age structure. The favourite is to subdivide the maximum plausible lifespan (or at least the maximum age for reproductive activity) into age classes of equal duration (x). The Leslie matrix[4,5,8] then relates the subpopulations of the various age classes at times t and t+x. The restriction to equal age intervals is the source of the mathematical simplicity of this approach but inevitably there is a price to pay; we cannot identify the age classes in the models with experimentally identifiable developmental stages (insect instars, cell cycle phases or whatever), a weakness which depending on the species being modelled, may be a mere computational irritant or a fatal flaw.

The continuous-time analogue of the Leslie matrix approach involves a partial differential equation[6,7,8] (the MacKendrick-von Foerster equation)

$$\frac{\partial f}{\partial t} + \frac{\partial f}{\partial a} + \delta(a,t)\, f = 0 \qquad (1)$$

describing the processes of aging and death, coupled to a renewal condition

$$R(t) = f(0,t) = \int_0^{\infty} \beta(a,t) f(a,t)\, da \qquad (2)$$

representing recruitment to the population. (The t-dependence in the functions δ and β covers both explicit time-dependence due to a changing environment and implicit time dependence representing density-dependent effects). The formal solution to these equations is well-known, namely

$$f(a,t) = f(0,t-a)\exp\{-\int_0^a \delta(x,t-a+x)dx\}, \qquad (3)$$

and is readily interpreted as the product of terms representing recruitment and survival. This solution does not take account of the presence in the population of remnants of the "initial age distribution" $f(a,0)$ if this is assumed arbitrary. If however we choose to complete the specification of the problem in terms of an <u>initial history</u> ($f(0,t)$ for all $t \leq 0$)) then it is complete provided the assumed history is a valid solution of equations (1) and (2).

This continuous-time approach shares one disadvantage with the Leslie matrix approach - the density function $f(a,t)$ is never an experimentally accessible quantity. This, however is only an irritant, the main weakness of the approach is its mathematical and computational intractability in situations with any serious density-dependence. (I still squirm with embarrassment at a summer spent generating artefactual "vole cycles" with the aid of equations (1) and (2) plus an unstable algorithm used uncritically!)

The recent work in our group has been directed to finding a compromise between the elegant but intractable von Foerster equation approach and the mathematically simple but biologically inadequate route using Leslie matrices. Our key simplifying assumptions are:

(a) We can identify, within the population, <u>subpopulations</u> of <u>functionally identical individuals</u> whose vital rates can be considered identical (e.g. insect larval instars).

(b) Transitions between categories (sub-populations) are triggered by a single factor (e.g. age, size, body weight, quota of limiting nutrient).

If $N_i(t)$ denotes the size of the i'th subpopulation at time t, then in a closed system (no immigration or emigration), the changes in N_i will be describable by an equation of the form

$$\dot{N}_i = \text{Recruitment} - \text{Maturation} - \text{Deaths}, \qquad (4)$$

the aim of the analysis being to derive expressions for the recruitment, maturation and death rates without invoking the detailed age-structure dynamics of the various subpopulations. Our progress to date is outlined in the next two sections.

3. AGE-DEPENDENT TRANSITIONS BETWEEN DEVELOPMENTAL STAGES

Having assumed that we may approximate the life history of a species as a series of developmental stages within which all individuals are functionally identical, we now require to specify the "single factor" responsible for triggering transitions between instars. The natural first candidate is age, so we now assume that transitions occur at fixed ages and derive expressions for the recruitment, maturation and death terms in equation (4). We have published the detailed algebra elsewhere[3], but the main results can be established intuitively using a "conveyer-belt" analogy, due originally I think to George Oster and illustrated in Fig. 2, whose application to this problem has been published elsewhere by us[17].

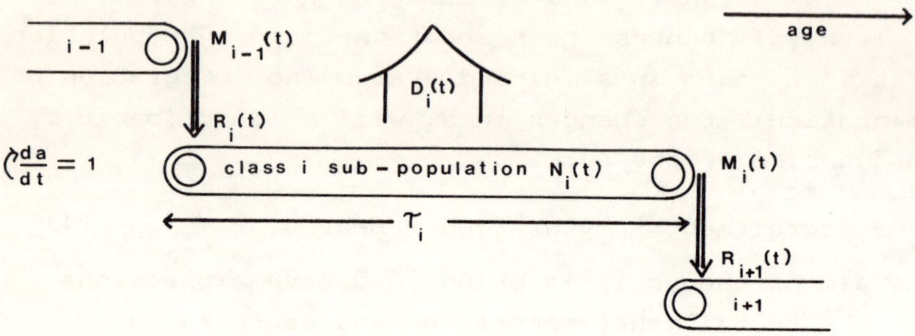

Fig. 2: An analogy of the dynamics of a single stage (not the first). The belt moves uniformly at unit speed.

We find that the <u>dynamics</u> of the i'th subpopulation are described by the pair of differential equations

$$\dot{N}_i(t) = R_i(t)-R_i(t-\tau_i)P_i(t)-\delta_i(t)N_i(t), \quad (5)$$

$$\dot{P}_i(t) = P_i(t)[\delta_i(t-\tau_i)-\delta_i(t)]. \quad (6)$$

in which

$$R_i(t) = R_{i-1}(t-\tau_i)P_i(t), \qquad i > 1 \quad (7)$$

$$R_1(t) = \Sigma_i\beta_i(t)N_i(t). \quad (8)$$

These equations can easily be integrated numerically (we shall demonstrate a package[9] using a simple predictor-corrector method later in the workshop) <u>provided we specify a valid initial history</u>. As the derivation of equations (5) - (8) involved the special solution (3), this is a far from trivial matter; there is however a consistent recipe which is particularly appropriate to laboratory populations.

We argue that most experiments start with an empty system into which individuals are added to start the experiment (at t=0 say). The representation of an empty system, namely

$$N_i(t) = 0, \quad R_i(t) = 0 \quad \text{for all } t<0 \tag{9}$$

is automatically a solution of equation (5) and is thus an acceptable "initial history". Our procedure is to adopt this empty state as the initial history and then carefully model the process of innoculation. We cannot retain the present "lumped" description <u>and</u> allow ourselves to add individuals of arbitrary age to the population at time t. However if we assume that all individuals added to an age class are newly qualified for that class, then we can retain equations (5) and (6) as the basic dynamic equations if we modify the definitions (7) and (8) to read

$$R_i(t) = R_{i-1}(t-\tau_i)P_i(t) + I_i(t), \tag{10}$$

$$R_1(t) = \Sigma_i \beta_i(t)N_i(t) + I_1(t). \tag{11}$$

Equations (5), (6), (10) and (11) constitute the basic description for lumped age-class models with transitions triggered by age.

To illustrate the use of the recipe, let me construct a very simple model of Lawton's experiments (Fig. 1(b)) on populations of the Indian meal-moth <u>*Plodia interpunctella*</u>. Adult <u>*Plodia*</u> do not feed, so the population was regulated by competition among larvae for food. We model this population by assuming that we need only consider two dynamically significant age classes:

(a) reproductively active adults of population $N_A(t)$

(b) a reproductively inactive larval stage of duration τ_L and population $N_L(t)$.

We assume a constant sex-ratio and that all females have a constant fecundity so that the net per capita adult fecundity is a constant (denoted by β_A), and we assume a constant, density-independent per capita adult death rate δ_A. We incorporate the resource limitation at the larval stage by assuming that a given larva competes equally with all other larvae (the "uniform competition" approximation[3]) so that the per capita death rate of the larvae is a function of $N_L(t)$. For mathematical simplicity we choose

$$\delta_L(t) = \alpha N_L(t), \tag{12}$$

where α is a constant. We can now refer to the recipe in equations (5), (6), (10) and (11) and deduce that the dynamics of the system are described by three delay-differential equations

$$\dot{N}_L = R_L(t) - R_L(t-\tau_L)P_L(t) - \alpha\, N_L^2(t) \tag{13}$$

$$\dot{N}_A = R_A(t) - \delta_A N_A(t) \tag{14}$$

with

$$\dot{P}_L(t) = \alpha P_L(t)[N_L(t-\tau_L) - N_L(t)] \tag{15}$$

where, if the system is innoculated by adding a few eggs (new born "larvae") just after t = 0 at a rate I(t),

$$R_L(t) = \beta_A N_A(t) + I(t); \quad R_A(t) = R_L(t-\tau_L)P_L(t) \qquad (16)$$

Fig. 3 contains a "typical" solution of equations (13)-(16) with parameters roughly appropriate to Lawton's Plodia - the main feature of the solution is a series of damped single-generation cycles with a period around 33 days. There is thus a possibility of explaining Lawton's observations as endogenous resonant quasi-cycles[8], the "driving" noise being demographic stochasticity.

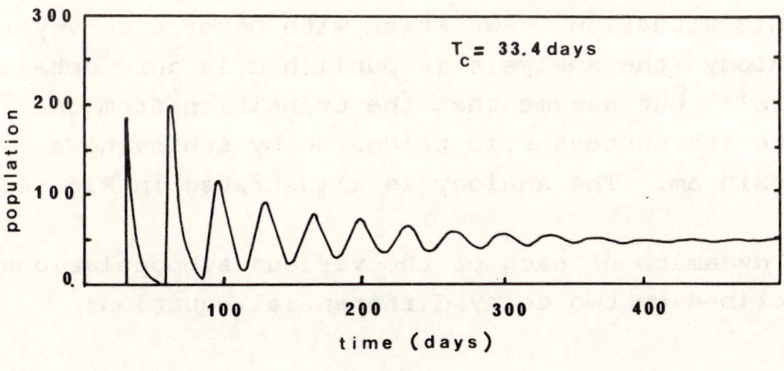

Fig.3: Damped "single-generation cycles" in a numerical solution of equs. (13)-(16). Details of the computation are in ref. 3.

4. GROWTH DEPENDENT TRANSITIONS

From physiological studies it is well known that for most insect species, it is not chronological age but weight gain that triggers the moult from one instar to its successor, a doubling of weight during an instar being typical (Dyar's "law"). There are also many studies which demonstrate linear or near-linear relationships between growth rate (weight gain per unit time) and food absorption rate (see e.g. Beddington, Lawton and Hassell[10]. It follows that in many situations the duration of any larval instar will itself vary dynamically in response to the availability of food.

This variability in instar duration must obviously infuence the population dynamics of the species concerned, yet there are scarcely any age-structure models which take explicit account of it. The formalism outlined in the previous section now begins to earn its keep as it is readily extended to cover this situation. We stick with Oster's conveyor belt analogy (the analysis is published is gory detail elsewhere[11]) but assume that the transition from one instar to its successor is triggered by achieving a weight gain Δm. The analogy is illustrated in Fig. 4.

The dynamics of each of the various subpopulations are described by two delay differential equations

$$\dot{N}_i(t) = R_i(t) - R_i(t-\tau_i(t)) \frac{g_i(t)}{g_i(t-\tau_i(t))} P_i(t) - \delta_i(t) N_i(t) \tag{17}$$

$$\dot{P}_i(t) = P_i(t) \left[\frac{g_i(t)\delta_i(t-\tau_i(t))}{g_i(t-\tau_i(t))} \quad \delta_i(t) \right] \tag{18}$$

with the evaluation of the delay given by a third, namely,

$$\tau_i(t) = 1 - g_i(t)/g_i(t-\tau_i(t)) \tag{19}$$

from which we can compute the lag at all times $t > 0$ if we know the value τ_{i0} of the lag at $t = 0$ and the "history" of the growth function at times $t \leqslant 0$. We can consistently interpret $g_i(t)$ at these times as the growth rate of a single individual introduced to the system at time t (< 0) and allowed to grow unimpeded by competition. This "unimpeded" growth rate will in many situations be constant (say g_{i0}); thus

$$\Delta m_i \equiv m_{i+1} - m_i = \int_{t-\tau_{i0}}^{t} g_i(t')dt' = g_{i0}\,\tau_{i0} \tag{20}$$

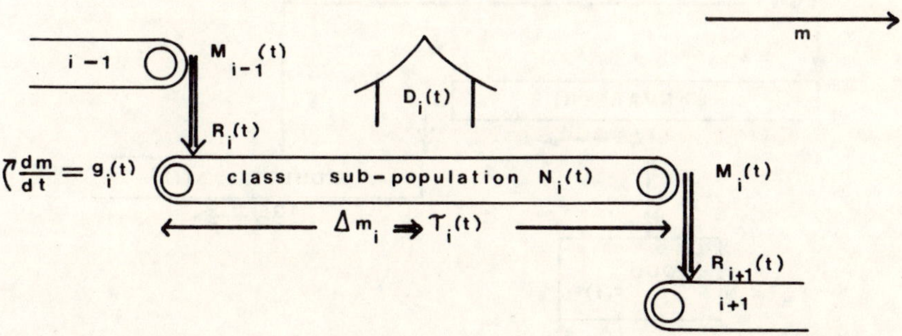

Fig.4: Transitions triggered by weight gain

We model the initial history as before and assume an empty system which a few individuals are added after t = 0. Thus equations (17) - (19) must be solved with

$$R_i(t) = R_{i-1}(t-\tau_{i-1}(t)) \frac{g_{i-1}(t)}{g_{i-1}(t-\tau(t))} P_{i-1}(t) + I_i(t), \quad i>1 \qquad (21)$$

$$R_i(t) = \sum_i \beta_i(t) N_i(t) + I_i(t), \qquad (22)$$

with $N_i(t) = 0$ for all $t \leq 0$, with the initial lag obtained from equation (20), and with

$$P_i(0) = \exp\{- \int_{-\tau_{i0}}^{0} \delta_i(t')dt'\} \qquad (23)$$

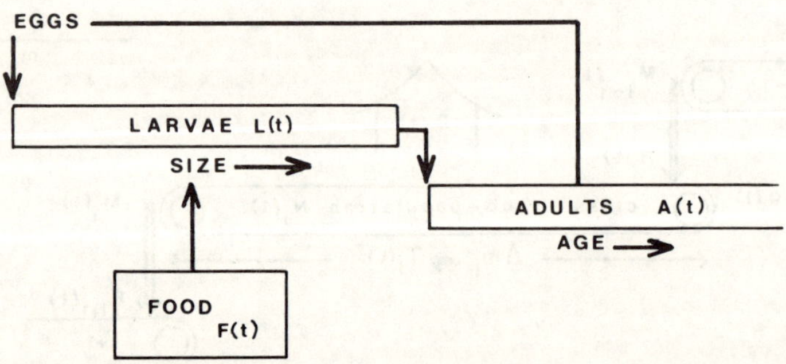

Fig.5 The "damselfly" model.

To illustrate how simple this recipe can be to use I shall construct another ludicrous caricature of larval competition - this one inspired by some experiments by Lawton, Thompson and Thompson[12] on the damselfly *Ischnura elegens* (van der Lind), certain instars of which have the ability to survive for long periods with no food, yet without a high level of deaths through starvation. Instars 10 and 12 could vary in length from 20 to 100 days with a high proportion of successful emergences at the end of the instar. The model is not a representation of a damselfly population, but solely a theoretical construct intended to elucidate a regulatory mechanism suggested by the damselfly experiments. The main features of the model are summarised in Fig. 5, the dynamics being describable by 4 differential equations

$$\dot{F}(t) = \Phi - N_L(t)A(F(t)) \tag{24}$$

$$\dot{N}_L(t) = R_L(t) - \frac{A(t)}{A(t-\tau_L(t))} \cdot R(t-\tau_L(t))P(t)-\delta_L N_L(t) \tag{25}$$

$$\dot{N}_A(t) = \frac{(A(t))}{A(t-\tau(t))} R_L(t-\tau_L(t))P(t) - \delta_A N_A(t) \tag{26}$$

$$\dot{\tau}(t) = 1 - \frac{(A(F(t))}{A(F(t-\tau_L(t))} \tag{27}$$

in which $F(t)$ represents the density of larval food which is supplied at a constant rate Φ, δ_L and δ_A denote the (constant) per capita death rates of larvae and adults respectively, $R_L(t)$ denotes the rate of recruitment to the larval population (the rate of production of viable eggs) and $A(t)$ the rate of absorption of larval food by a single larva. We assume adult fecundity to be density independent so that

$$R_L(t) = \beta\, N_A(t) \tag{28}$$

and we assume a Michaelis-Menten feeding function implying

$$A(t) = \frac{A_{max}F(t)}{K_F + F(t)} \tag{29}$$

The quantity $\tau_L(t)$, represents the time spent as larvae by individuals who mature to adulthood at time t, so the through-age-class survival function for the larvae is

$$P(t) = \exp\{- \delta_L \tau_L(t)\} \tag{30}$$

The model suggests the possibility that population size may be regulated solely by changes in instar duration, and numerical solutions of the system described by equations (24) - (30) confirm this. The results have been published elsewhere[11], but we here reproduce one set of solutions (Fig. 6), which, illustrate clearly the adjustments to the lag which produce the stabilization.

5. FURTHER DEVELOPMENTS

This paper has outlined a framework within which age - and/or size-structured populations can be modelled using delay-differential equations. There are still many outstanding problems requiring solution before we have an adequate tool-kit for modelling experiments (like those of Lawton on *Plodia*) involving larval competition.

The most obvious gap in the present formalism is a representation of random variation in instar duration (as opposed to the systematic variation described in the previous section). Experiments by Podoler[13] on *Plodia* suggest that this random variation is as important as the systematic "stretching" of the larval stage induced by starvation and there are many similar well-documented examples (ref. 14 and references therein). Blythe[14,15] has made a start to this problem by developing a rigorous procedure applicable to insect populations with *adult* competition only;

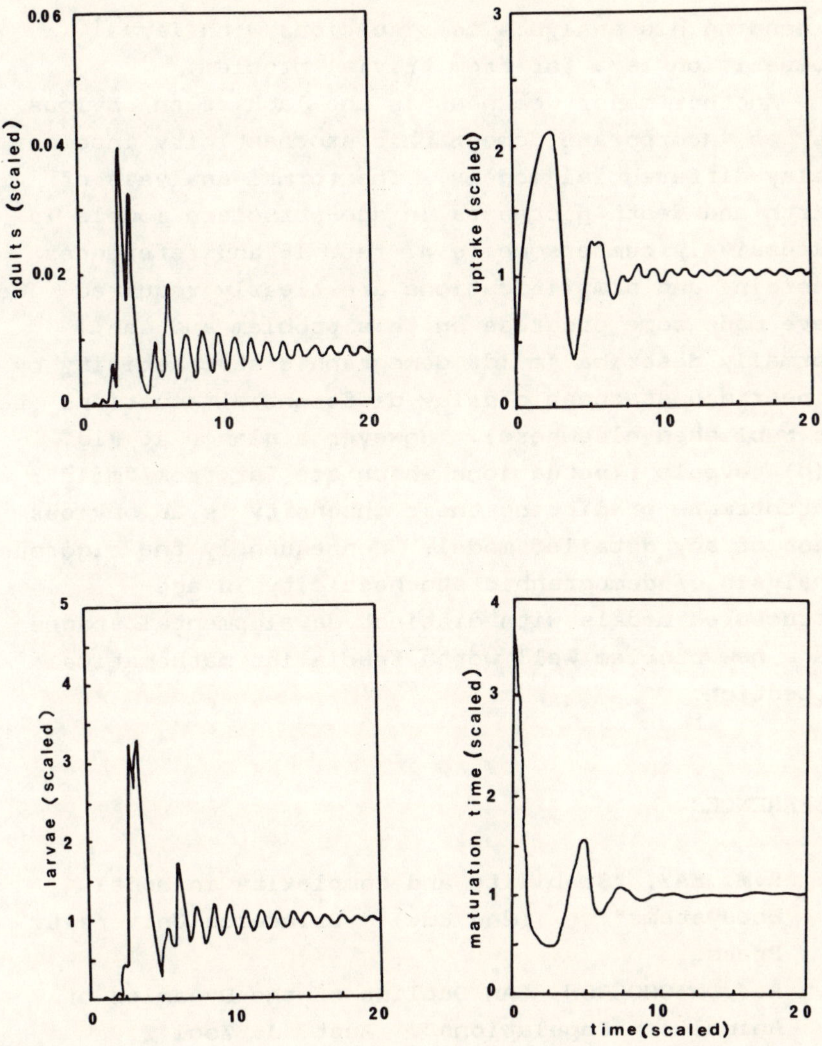

Fig.6: One set of solutions for the "damselfly" model illustrating population regulation mediated by a variable maturation time.

extending his analysis to situations with larval competition is a far from trivial problem.

Another major weakness is the lack of an obvious way to incorporate demographic stochasticity into our delay-differential models. The formal analysis of birth and death processes in age-structure models is excessively cumbersome (e.g. ref. 16 and references therein) and simplifications are clearly required. We have made some progress on this problem and can formally describe "mild" demographic stochasticity by linearized stochastic delay-differential equations (to be published elsewhere). However a glance at Fig. 1(b) reveals fluctuations which are far from "mild"; furthermore predicting their intensity is an obvious test of any detailed model. Consequently the rigorous analysis of demographic stochasticity in age-structured models with distinct developmental stages is a new problem well worth specialist mathematical attention.

REFERENCES

1. R.M. MAY, "Stability and Complexity in Model Ecosystems" (2nd Ed.). Princeton University Press.
2. A.J. NICHOLSON, "An Outline of the Dynamics of Animal Populations". Aust. J. Zool 2, 9-65 (1954).
3. W.S.C. GURNEY, R.M. NISBET, J.H. LAWTON, "The Systematic Formulation of Tractable Single-Species Population Models Incorporating Age-Structure". J. Anim. Ecol. - to appear 1983.
4. P.H. LESLIE, "On the Use of Matrices in Certain Population Mathematics". Biometrika 33, 183-212 (1945).

5. P.H. LESLIE, "Some Further Notes on the Use of Matrices in Population Mathematics". Biometrika 35, 213-245 (1948).
6. H. Von FOERSTER, "Some Remarks on Changing Populations" in "The Kinetics of Cellular Proliferation" (F. Stohlman, Jr.Ed.) p. 382-407. Grune and Stratton, New York, 1959.
7. A.E. MacKENDRICK, "Applications of Mathematics to Medical Problems". Proc. Edin. Math. Soc. Soc. 44, 98-130 (1926).
8. R.M. NISBET and W.S.C. GURNEY, "Modelling Fluctuating Populations", John Wiley and Sons Ltd. (1982).
9. P. MAAS, R.M. NISBET and W.S.C. GURNEY, (1982) "Solver: an Adaptable Program Template for Initial Value Problem Solving". Applied Physics Industrial Consultants, University of Strathclyde, Glasgow, U.K.
10. J.R. BEDDINGTON, J.H. LAWTON and M.P. HASSELL, "The Components of Arthropod Predation, II: The Predator Rate of Increase". J. Anim. Ecol. 45, 165-185 (1976).
11. R.M. NISBET and W.S.C. GURNEY, "The Systematic Formulation of Population Models for Insects with Dynamically Varying Instar Duration". Theor. Pop. Biol. - to appear (1983).
12. J.H. LAWTON, B.A. THOMPSON and D.J. THOMPSON, "The Effects of Prey Density on the Survival and Growth of Damselfly Larvae". Ecological Entomology 5, 39-51.(1980)
13. H. PODOLER. "Effects of Intraspecific Competition in the Indian Meal-Moth Plodia interpunctella (Hubner) (Lepidoptera:Physidiae) on Populations of the Moth and its Parasite Nemeritis Canesions (Gravenhurst) (Hymenoptera: Icheumonidore)", J. Anim. Ecol.

43, 641-651 (1974).

14. S.P. BLYTHE. "Simple Age-Structure Models of Laboratory Insect Populations" PhD Thesis, University of Strathclyde, 1982.
15. S.P. BLYTHE, R.M. NISBET and W.S.C. GURNEY, "Formulating Population Models with Differential Ageing". Submitted to "Proceedings of International Conference on Population Biology - Edmonton 1982". Springer-Verlag, to appear 1983
16. J.H. POLLARD, "Mathematical Models for the Growth of Human Populations". Cambridge University Press 1973.
17. W.S.C. GURNEY and R.M. NISBET, "The Systematic Formulation of Delay-differential Models of Age or Size Structured Populations". Submitted to "Proceedings of International Conference on Population Biology - Edmonton 1982". Springer-Verlag - to appear 1983.
18. R.M. NISBET and W.S.C. GURNEY, "Models of Uniform Larval Competition", Submitted to "Procedings of the Autumn Course in Mathematical Ecology" (S.A. Levin, Editor). Springer-Verlag, to appear 1983.

R.M. Nisbet and W.S.C.Gurney
Depaartment of Applied Physics
University of Strathclyde
Glasgow G4 ONG
U.K.

The FitzHugh-Nagumo equations: a simplified model of nerve impulse transmission

B.D. SLEEMAN

ABSTRACT

The classic Hodgkin-Huxley system of differential equations governing nerve impulse transmissions presents a major challenge to the Mathematical Analyst, although there have been a number of numerical experiments performed in these equations. In the early 1960's FitzHugh and Nagumo proposed a simpler model based on the Van-der-Pol oscillator. This system is more tractable than the H-H system; it preserves many of the features of the H-H system and is more amenable to analytical and numerical treatment. In this paper we shall review some aspects of the state of knowledge regarding the FitzHugh-Nagumo system. Furthermore, under a further simplying assumption, we shall demonstrate how the system may be solved by quite elementary means. This in turn will reveal some new and unexpected properties of the system which are open to physiological interpretation. Finally we shall indicate some open problems and new avenues of research.

§1 INTRODUCTION - THE FITZHUGH-NAGUMO SYSTEM AND RELATED MODELS

A widely accepted model used to describe the propagation of stimuli in an unmylenated nerve axon is the following system of four non-linear differential equations developed by Hodgkin and Huxley (9).

$$\begin{aligned} u_t &= u_{xx} - F(u,m,n,h), \\ m_t &= \gamma_m(u)\big(m_\infty(u) - m\big), \\ n_t &= \gamma_n(u)\big(n_\infty(u) - n\big), \\ h_t &= \gamma_h(u)\big(h_\infty(u) - h\big), \end{aligned} \tag{1}$$

where $x \in (-\infty,\infty)$ denotes distance along a single nerve axon t is time, u is the potential difference across the nerve membrane, m denotes sodium ion activation, n denotes potassium ion activation and h denotes sodium ion deactivation.

If, as is commonly performed in experiments, a segment of axon is held at a fixed current value I then we have the 'space-clamp' model and the first of equations (1) takes the form

$$u_t = I - F(u,m,n,h) \tag{2}$$

where I is the applied current. In equations (1), (2) the reactive term F is a complicated function of its argument as are the functions $\gamma_\alpha(u)$, $\alpha = m,n,h$. From the analytical point of view (1) is very difficult to analyse and has yielded few of its secrets.

In order to reduce the complexity of the problem one is led to consider alternative models to (1) which preserve qualitatively the main features of impulse transmission predicted by (1) but which is amenable to analytic treatment.

Such a model is available; is due to FitzHugh (7) and Nagumo et al (20) and is based on the Van-der-Pol oscillator. It can be written in the form

$$\begin{aligned} u_t &= u_{xx} + f(u) - v \\ v_t &= \sigma u - \gamma v , \end{aligned} \tag{3}$$

or in the space clamp case as,

$$\begin{aligned} u_t &= I + f(u) - v , \\ v_t &= \sigma u - \gamma v . \end{aligned} \tag{4}$$

In these models $\sigma,\gamma > 0$, $0 < a < 1$, are constants and $f(u) \equiv u(1-u)(u-a)$. The interpretation of the various quantities in (3), (4) are

u - membrane potential

$f(u)$ - sodium ion activation

v - potassium ion activation

and I - applied current.

Note that in the FitzHugh-Nagumo system there is no counterpart to sodium ion inactivation.

Models (3) and (4) do yield to mathematical analysis and are used to suggest properties of (1) and of nerve physiological phenomena in general.

Since (3), (4) are proposed to represent the qualitative behaviour of nerves we can reduce the complexity still further by replacing the cubic polynomial f by the piecewise linear function

$$F(u) = -u \ , \quad u < \frac{a \tan \theta}{1 + \tan \theta} \ ,$$

$$= \tan \theta (u-a) \ , \ \frac{a \tan \theta}{1 + \tan \theta} \leq u \leq \frac{1 + a \tan \theta}{1 + \tan \theta} \ , \tag{5}$$

$$= 1 - u \ , \quad u > \frac{1 + a \tan \theta}{1 + \tan \theta} \ .$$

In this case models (3), (4) yield to "linear" analysis and offer the possibility of obtaining exact analytic solutions to problems. This last simplification is due to A C Scott and has been exploited by McKean (18), Rinzel and Keller (24) and others.

In summary then we have the following degrees of complexity

A) Piecewise-linear FitzHugh-Nagumo Systems
(Linear Analysis)

↓

B) FitzHugh-Nagumo System
(Non-linear Analysis)

↓

C) Hodgkin-Huxley System
(Non-linear Analysis)

↓

D) Physiological interpretation and experiment.

For the most part we shall concentrate on A, B and D, and although much is now known about the FitzHugh-Nagumo equations we shall direct attention to only the more recent developments, indicate some unusual properties and point to areas where new and unexpected phenomena are likely to appear.

To conclude this introduction we mention two related model systems of equations which are also amenable to some of the ideas proposed in this paper.

Zeeman's Heart Beat Equations:-

$$\varepsilon \dot{x} = -(x^3 - Tx + b) \ ,$$
$$\dot{b} = (x - x_0) \ , \tag{6}$$

where

ε - a scaling parameter

x - a typical muscle fibre length

b - electrochemical control variable

T - tension

and x_0 - position of diastole.

Belousov-Zhabotinskii chemical reaction

$$x_t = u_{xx} + u(1-u-rv)$$
$$v_t = v_{xx} - buv \quad . \tag{7}$$

For a general introduction to the models described here we cite Jones and Sleeman (12).

§2 AN INITIAL VALUE PROBLEM

Suppose the axon occupies the interval $0 \le x < \infty$ and is stimulated by a pulse $P(t)$ at $x = 0$, then the problem to be solved is

$$\begin{aligned} u_t &= u_{xx} + f(u) - v \; , \\ v_t &= \sigma u - \gamma v \; , \qquad x,t \in (0,\infty) \times (0,\infty) \end{aligned}$$

with (8)

$$\begin{aligned} u(x,0) &= 0 \\ u(0,t) &= P(t) \; . \end{aligned}$$

This problem has been considered by Rauch-Smoller (23), who prove that (8) has a unique solution which depends continuously on $P(t)$ and the parameters a, σ, γ. Thus the problem is "well-posed". The proofs are based on the concept of "invariant rectangles" as developed by Conley (5), Chueh et al (4). See also the work of Aronson and Weinberger (1).

An important question of course is:
How does the solution to (8) under the stimulus $P(t)$ evolve in time? We shall return to this question later.

Among the many important solutions to the FitzHugh-Nagumo equations the study of travelling waves occupies a central position. Travelling waves are solutions to the governing equations (8) of the form

$$u(x,t) = \phi(x+ct) \; , \quad v(x,t) = \psi(x+ct) \tag{9}$$

where c is the wave speed and $\phi(\xi)$, $\psi(\xi)$ $(\xi = x+ct)$ satisfy the conditions

$$\phi,\quad \phi',\quad \psi \to 0 \quad \text{as} \quad |\xi| \to \infty\,. \tag{10}$$

From (8) it follows that travelling waves solve the first order system

$$\begin{aligned} \eta' &= c\eta + f(\phi) - \psi\,, \\ \phi' &= \eta\,, \\ \psi' &= \frac{\sigma}{c}\phi - \frac{\gamma}{c}\psi\,. \end{aligned} \tag{11}$$

This system has been considered at length by Hastings (9), Carpenter (2), Conley (6) and others to establish existence of homoclinic orbits, bursting phenomena, periodic solutions, etc. In addition Green and Sleeman (8) and Sleeman (27) have used simple calculus methods to obtain qualitative properties of travelling waves. First of all the system (11) has $(0,0,0)$ as the unique rest state if and only if

$$\frac{\sigma}{\gamma} > \frac{(1-a)^2}{4}\,.$$

It is also proved that travelling waves exist if and only if $0 < a < \frac{1}{2}$. Furthermore it follows from the existence theory and numerical evidence that there are two possible speeds of propagation for the travelling wave, the faster of which is believed to be stable (see below) and has a speed satisfying the inequality

$$c^2 > \frac{4\sigma}{(1-a)^2} > \gamma\,.$$

It turns out that 'a' can be thought of as a threshold parameter in that for ϕ to be a travelling wave we must have

$$\phi_{max} > a\,.$$

The computation of travelling waves is a delicate matter and has been considered by several authors; the most recent of which are the comprehensive results of Miura (19). However if we consider the 'piecewise linear' simplification (5) then it is possible to construct analytic forms for the travelling wave. This approach has been exploited to great effect by McKean (18), Rinzel and Keller (24) and others. Details of the method are to be found in Jones and Sleeman (12) and will be developed in a wider context in the following section.

§3 EVOLUTION OF TRAVELLING WAVES

In practice one rarely observes travelling waves:- they usually evolve as 'asymptotic states' ($t \to \infty$) from the system (8). It is thus a fundamental problem to examine the asymptotic behaviour of solutions to (8) with time. In Levine and Sleeman (15) it is proved that if $|P(t)| < a$ for all t then $\{u(x,t), v(x,t)\} \to 0$, as $t \to \infty$, exponentially. This result provides further evidence that 'a' is a threshold parameter. A more detailed analysis of the characteristics required of $P(t)$ in order for travelling waves to evolve have been considered by Schonbek (26).

Intimately related to the problem of evolution of travelling waves is their stability. It has been shown by Sleeman (28) that the 'slower' wave is unstable. The stability of the 'faster' wave is a much more difficult problem and has only very recently been established by Jones (11).

The ideal of course would be to attempt an analytic solution to problem (8). This is clearly impossible in general, although there are good numerical results available as reported by Manoranjan (17) in a companion paper to this volume. If however we use the caricature (5) then it is possible to obtain analytic solutions. This programme has been carried out with the author's student E Tuma and we choose to illustrate the idea in the following simpler problem arising in population genetics, viz

$$\begin{aligned} u_t &= u_{xx} + G(u)\ , \quad x \in \mathbb{R}, \quad t > 0 \\ u(x,0) &= g(x) \end{aligned} \tag{12}$$

where $g(x)$ is a monotone decreasing function of x, i.e. $g(\infty) = 0$, $g(-\infty) = 1$ and

$$\begin{aligned} G(u) &= u \tan\alpha\ , \quad u < \frac{\tan\beta}{\tan\alpha + \tan\beta}\ , \\ &= \tan\beta(1-u)\ , \ u > \frac{\tan\beta}{\tan\alpha + \tan\beta}\ , \ (\alpha,\beta) \in (0,\pi/2). \end{aligned} \tag{13}$$

In particular we shall seek a solution to (13) for $0 < u < 1$ with

$$g(x) = \frac{\tan\beta}{\tan\alpha + \tan\beta}(a_1 e^{-\alpha_1 x} + a_2 e^{-\alpha_2 x}) \equiv g_1(x)\ , \quad x > 0\ ,$$

$$= \frac{\tan\beta}{\tan\alpha + \tan\beta} \quad , \quad x = 0 \; ,$$

$$= 1 - \left(1 - \frac{\tan\beta}{\tan\alpha + \tan\beta}\right) e^{\beta_1 x} \equiv g_2(x), \; x < 0, \tag{14}$$

where $a_1 + a_2 = 1$.

The procedure is to decompose the problem (12) into the two problems

$$\begin{aligned} \text{I} : \; & u_t = u_{xx} + u \tan\alpha \, , \quad x > 0 \, , \quad t > 0 \\ & u(x,0) = g_1(x) \, , \end{aligned} \tag{15}$$

$$\begin{aligned} \text{II} : \; & v_t = v_{xx} + \tan\beta(1-u) \quad x < 0, \quad t > 0 \\ & v(x,0) = g_2(x) \; . \end{aligned} \tag{16}$$

The important part of the technique is to 'match' $u(x,t)$, $v(x,t)$ along a curve

$$x = \gamma(t) \; , \; \gamma(0) = 0$$

$$\text{such that} \quad u\big(\gamma(t),t\big) = v\big(\gamma(t),t\big) = \frac{\tan\beta}{\tan\alpha + \tan\beta} \; . \tag{17}$$

If we make the changes of variable

$$z_1 = x - \gamma_1(t) \; , \quad z_2 = \gamma_2(t) - x \tag{18}$$

then problems I, II can be replaced by the equivalent problems

$$\begin{aligned} \text{I}' \; : \; & u_t = u_{z_1 z_1} + \gamma_1'(t) u_{z_1} + \tan\alpha \, u \, , \quad z_1 > 0, \; t > 0 \\ & u(z_1,0) = g_1(z_1) \, , \quad z_1 > 0 \\ & u(0,t) = \frac{\tan\beta}{\tan\alpha + \tan\beta} \, , \quad t > 0 \end{aligned} \tag{19}$$

$$\begin{aligned} \text{II}' \; : \; & v_t = v_{z_1 z_2} - \gamma_2'(t) v_{z_2} + \tan\beta(1-u) \\ & v(z_2,0) = g_2(z_2) \\ & v(0,t) = \frac{\tan\beta}{\tan\alpha + \tan\beta} \, , \quad t > 0 \; . \end{aligned} \tag{20}$$

These two problems are then to be linked by finding a function $\gamma(t)$ such that

$$\gamma_1(t) = \gamma_2(t) = \gamma(t) \quad \text{for all} \quad t > 0.$$

Problem I' can be formally solved by writing $u(z_1,t)$ in the form

$$u(z_1,t) = \frac{1}{2\pi i}\int_\Gamma A(\beta)e^{(\beta^2+\tan\alpha)t + \gamma_1(t)\beta + \beta z_1}\,d\beta$$

where Γ is a suitably chosen contour in the complex β-plane and the function $A(\beta)$ is to be determined.

From the initial and boundary conditions we require

$$u(z_1,0) = g_1(z_1) = \frac{1}{2\pi i}\int_\Gamma A(\beta)e^{\beta z_1}\,d\beta\,, \tag{21a}$$

$$u(0,t) = \frac{\tan\beta}{\tan\alpha+\tan\beta} = \frac{1}{2\pi i}\int_\Gamma A(\beta)e^{(\beta^2+\tan\alpha)t + \gamma_1(t)\beta}\,d\beta. \tag{21b}$$

In (21a) $g_1(z_1)$ can be thought of as the inverse Laplace transform of $A(\beta)$. This then determines Γ as well as the unknown function $A(\beta)$. A residue calculation is then used in (21b) to determine the function $\gamma_1(t)$. The same idea is used for problem (19) to provide $\gamma_2(t)$.

From (21a) we find that

$$A(\beta) = \frac{\tan\beta}{\tan\alpha + \tan\beta}\left(\frac{a_1}{\beta+\alpha_1} + \frac{a_2}{\beta+\alpha_2}\right),$$

and from (21b) it follows that

$$a_1e^{(\alpha_1^2+\tan\alpha)t-\gamma_1(t)\alpha_1} + a_2e^{(\alpha_2^2+\tan\alpha)t-\gamma_1(t)\alpha_2} = 1. \tag{22}$$

In order for (22) to be compatible it follows that we must have

$$\gamma_1(t) = \frac{(\alpha_1^2+\tan\alpha)t}{\alpha_1} = \frac{(\alpha_2^2+\tan\alpha)t}{\alpha_2}. \tag{23}$$

Thus in this particular example the exponents of the exponential functions occurring in the initial condition function $g_1(z)$ cannot be arbitrarily chosen by the above method of solution.

Proceeding along the same lines for problem II' we find that

$$\gamma_2(t) = \frac{-(\beta_1^2-\tan\beta)t}{\beta_1}. \tag{24}$$

There are now two cases to consider.

Case 1

Suppose $\alpha_1 = \alpha_2 = \bar{\alpha}$ then $\gamma_1(t) = \frac{(\bar{\alpha}^2 + \tan\alpha)t}{\bar{\alpha}}$ and for $\gamma_1(t) = \gamma_2(t)$ we must have

$$\frac{\bar{\alpha}^2 + \tan\alpha}{\bar{\alpha}} = -\frac{(\beta_1^2 - \tan\beta)}{\beta_1} \;. \tag{25}$$

In other words

$$\begin{aligned} g(x) &= \frac{\tan\beta}{\tan\alpha + \tan\beta}\, e^{-\bar{\alpha}x} \;, \quad x > 0 \;, \\ &= \frac{\tan\beta}{\tan\alpha + \tan\beta} \;, \quad x = 0 \;, \\ &= 1 - \left(1 - \frac{\tan\beta}{\tan\alpha + \tan\beta}\right) e^{\beta_1 x} \;, \quad x < 0 \end{aligned}$$

where $\bar{\alpha}$ and β_1 are given as above, evolves as

$$\begin{aligned} u(x,t) &= \frac{\tan\beta}{\tan\alpha + \tan\beta}\, e^{\bar{\alpha}\{(\frac{\bar{\alpha}^2 + \tan\alpha}{\bar{\alpha}})t - x\}} \;, \quad x \geq \frac{\bar{\alpha}^2 + \tan\alpha\; t}{\bar{\alpha}} \;, \\ &= 1 - \left(1 - \frac{\tan\beta}{\tan\alpha + \tan\beta}\right) e^{\beta_1\{x - (\frac{\bar{\alpha}^2 + \tan\alpha}{\bar{\alpha}})t\}} \;, \quad x \leq \left(\frac{\bar{\alpha}^2 + \tan\alpha}{\bar{\alpha}}\right)t. \end{aligned} \tag{26}$$

Case 2

If $\alpha_1 \neq \alpha_2$ then $\alpha_2 = \frac{\tan\alpha}{\alpha_1}$ and

$$\frac{\beta_1^2 - \tan\beta}{\beta_1} = -\frac{(\alpha_1^2 + \tan\alpha)}{\alpha_1} \;.$$

In this case the initial data

$$\begin{aligned} g(x) &= \frac{\tan\beta}{\tan\alpha + \tan\beta}\left(a_1 e^{-\alpha_1 x} + a_2 e^{-\frac{\tan\alpha}{\alpha_1}x}\right) \;, \quad x > 0 \\ &= \frac{\tan\beta}{\tan\alpha + \tan\beta} \;, \quad x = 0 \\ &= 1 - \left(1 - \frac{\tan\beta}{\tan\alpha + \tan\beta}\right) e^{\beta_1 x} \;, \quad x < 0 \end{aligned}$$

evolves as

$$u(x,t) = \frac{\tan\beta}{\tan\alpha+\tan\beta}\{a_1 e^{(\alpha_1^2+\tan\alpha)t-\alpha_1 x}$$

$$+ a_2 e^{\left(\left(\frac{\tan\alpha}{\alpha_1}\right)^2+\tan\alpha\right)t - \frac{\tan\alpha}{\alpha_1}x}\} \quad x \geq \frac{(\alpha_1^2+\tan\alpha)}{\alpha_1} t \qquad (27)$$

$$= 1 - \left(1-\frac{\tan\beta}{\tan\alpha+\tan\beta}\right)e^{(\beta_1^2-\tan\beta)t+\beta_1 x}$$

$$x \leq \frac{(\alpha_1^2+\tan\alpha)t}{\alpha_1} .$$

It should be noted that the solutions (26), (27) to problems (12), (13), (14) are in agreement with the asymptotic results of Larson (13), Rothe (25), Uchiyama (32) and others. Furthermore the ideas outlined above have been applied to a wider class of initial data than (14); to problems where $G(u)$ in (12) is replaced by the piecewise linear form $F(u)$ defined by (5) and most importantly to the FitzHugh-Nagumo system. These results will be reported elsewhere.

§4 SMALL AMPLITUDE PERIODIC WAVES

Suppose we relax the condition that the FitzHugh-Nagumo system has a unique state. That is suppose $\frac{\sigma}{\gamma} < \frac{(1-a)^2}{4}$, then some interesting new features arise. For a piecewise linear model similar to (3), (5) Pauwelussen (22) has demonstrated the existence of various types of wave front solutions (i.e. heteroclinic orbits) linking the "inner" and "outer" rest states. More recently Sleeman (29) has demonstrated the existence of small amplitude periodic solutions oscillating about the "middle" rest. These solutions have, among other things, interesting stability properties.

If we consider the piecewise linear model (3), (5) then we establish the following.

Proposition 1

Let $\gamma < \frac{\sigma}{\gamma} < \frac{(1-a)\tan\theta}{1 + a\tan\theta}$, then the FitzHugh-Nagumo system (3), (5) has a 2-parameter family of small amplitude periodic solutions for a given period T of the form

$$u(kx+wt) \equiv u(\xi) = \frac{\gamma\ a \tan\theta}{(\gamma \tan\theta - \sigma)} + A \sin\frac{2\pi}{T}(\xi+\tau) \qquad (28)$$

for some A, τ and where

$$k^2 = \frac{(\tan\theta-\gamma)}{4\pi^2}T^2, \quad w^2 = \frac{(\sigma-\gamma^2)}{4\pi^2}T^2 .$$

The wave speed c is independent of T and is given by

$$c^2 = \frac{\sigma-\gamma^2}{\tan\theta-\gamma} .$$

Proposition 2

Let $u(x,t)$ be a solution of the FitzHugh-Nagumo system (3), (5) where the initial data u_0 is periodic of period T_0 and $u(x,t)$ is suitably bounded [see B D Sleeman (1982) for details].

Then (i) if $\frac{4\pi^2}{T_0^2} > \tan\theta-\gamma$, $u(x,t)$ is stable and represents a damped oscillation which tends to the middle rest state $u = \frac{a\,\gamma \tan\theta}{(\gamma \tan\theta-\sigma)}$ as $t \to \infty$.

(ii) if $\frac{4\pi^2}{T_0^2} = \tan\theta-\gamma$, $u(x,t)$ is neutrally stable. In particular the periodic solution of proposition 1 is neutrally stable.

(iii) if $\frac{4\pi^2}{T_0^2} < \tan\theta-\gamma$, $u(x,t)$ is locally unstable and may evolve into a wave front approaching the outer rest state $u = \frac{\gamma}{(\sigma+\gamma)}$ or approach zero as $t \to \infty$.

Which of the events in (iii) occurs depends on the position of the middle rest state relative to the inner and outer rest states. If the middle rest state is nearer the outer rest state then $u(x,t) \to 0$ as $t \to \infty$, while if it is nearer the inner rest state then $u(x,t)$ evolves as a front linking with the outer rest state.

We refer to Manoranjan (17) for numerical confirmation of these results.

§5 THE SPACE CLAMP CASE

Here we consider the model

$$u_t = I + f(u) - v ,$$
$$v_t = \sigma u - \gamma v , \tag{29}$$
$$\sigma, \gamma > 0 , \quad 0 < a < 1 .$$

This model is available for comparison with experimental work and so some of the phenomena predicted by (29) could be tested in the laboratory.

Suppose the current I is constant then the system (29) could be analysed in the Poincaré phase plane. It turns out that I can be treated as a bifurcation parameter and the problem can be analysed via the theory of Hopf bifurcation. Troy (31) has carried out such an analysis leading to some interesting and unexpected results.

If I is allowed to vary with time and in particular if I is a periodic function of t then some quite remarkable features appear which may have some important physiological consequences. In order to illustrate some of the phenomena to be expected consider the case in which $\gamma = 0$. First of all make the change of scale $t = \tau/\sqrt{\sigma}$ then on eliminating v from (29) we arrive at the equation

$$\sqrt{\sigma}\, u_{\tau\tau} - f'(u)u_\tau + \sqrt{\sigma}\, u = \dot{I}(\tau) . \tag{30}$$

This equation is of the general Van-der-Pol type

$$\varepsilon\, u'' + \phi(u)u' + \varepsilon\, u = b\, P(t) \tag{31}$$

considered by Cartwright and Littlewood (3) and Levinson (16). Recently Levi (14) has analysed problems of the form (31) using the bifurcation theories of Smale (30) and Newhouse and Palis (21) as well as the Poincaré map.

Here we list some of the results to be expected from (30) when $I(\tau) = b \sin \frac{2\pi t}{T}$, for some period T and parameter b. Details of this work will appear elsewhere.

There exists an interval of values of b say $[b_1, b_2]$ consisting of sub-intervals A_k, B_k separated by 'thin' gaps g_k (small compared with $\sqrt{\sigma}$) such that if we look at the Poincaré map $D : (u,u_\tau)_{\tau=0} \to (u,u_\tau)_{\tau=T}$ for some appropriate period T then

1) the qualitative behaviour of D throughout each interval A_k, B_k is preserved, while the g_k are bifurcation intervals.

2) for all $b \in [b_1, b_2]$, D has one totally unstable fixed point x_0.

3) for $b \in A_k$, D is of the so-called Morse-Smale type. That is D has exactly one pair of periodic points of period (2n-1) with $n = n(\sigma) \sim \frac{1}{\sqrt{\sigma}}$ an integer which is constant throughout each A_k.

   One of these points is a sink, the other is a saddle.

4) for $b \in B_k$ the invariant set of D consists, in addition to x_0, of two sink-saddle pairs of periods (2n+1), (2n-1) and an invariant Cantor set C (of measure zero) to which the saddles belong. The set C can be thought of as the set of points which are undecided as to which of the two sinks to tend for future iterates.

Physiologically these results mean that in the space clamp case it is possible to choose I(t) so that random or even chaotic repetitive firing can occur in the nerve axon.

When this analysis is applied to the heart beat equations (6) it could be used to explain certain irregular heart beat behaviour. Some pointers in this direction are to be found in the results recently reported by Winfree (33).

REFERENCES

(1) D G Aronson and H F Weinberger: Nonlinear diffusion in population genetics, combustion and nerve propagation. Proceedings of the Tulane Program in Partial Differential Equations, Lecture Notes in Mathematics, 446, Springer-Verlag 1975.

(2) G Carpenter: A geometric approach to singular perturbation problems with applications to nerve impulse equations. J. Differential Equations 23 (1977).

(3) M L Cartwright and J E Littlewood: On non-linear differential equations of the second order: I. The equation $\ddot{y} - k(1-y^2)\dot{y} + y = b\lambda k \cos(\lambda t+\alpha)$, k large. J. London Math. Soc. 20 (1945), 180-189.

(4) K N Chueh, C C Conley and J A Smoller: Positively invariant regions for systems of non-linear diffusion equations. Indiana Univ. Math. J. 26 (1977), 373-392.

(5) C C Conley: Isolated Invariant Sets and the Morse Index. CBMS - AMS Providence, R.I., 1978.

(6) C C Conley: On travelling wave solutions of non-linear diffusion equations. M.R.C. Tech. Summary Report 1492, University of Wisconsin, 1975.

(7) R FitzHugh: Mathematical models of excitation and propagation in nerve. Biological Engineering (H P Schwan Ed.) New York: McGraw-Hill 1969.

(8) M W Green and B D Sleeman: On FitzHugh's nerve axon equation. J. Math. Biol. 1 (1974), 153-163.

(9) S P Hastings: On the existence of homoclinic and periodic orbits for the FitzHugh-Nagumo equations. Quart. J. Math. (Oxford) 27 (1976), 123-134.

(10) A L Hodgkin and A F Huxley: A quantitative description of membrane current and its application to conduction and excitation in nerve. J. Physiol. 117 (1952), 500-544.

(11) C Jones: Stability for the impulse solution of the FitzHugh-Nagumo system. Abstracts - AMS 3 No 6, 82C-35-35 (1982).

(12) D S Jones and B D Sleeman: Differential Equations and Mathematical Biology, George Allen and Unwin, London (1983).

(13) D A Larson: Transient bounds and time-asymptotic behaviour of solutions to non-linear equations of Fisher type. SIAM J. Appl. Math. 34 (1978), 93-103.

(14) M Levi: Qualitative Analysis of the Periodically Forced Relaxation Oscillation, Memo - AMS 244 (1981).

(15) H A Levine and B D Sleeman: A note on the asymptotic behaviour to Nagumo's equations. Battelle Advanced Studies Centre, Geneva, Mathematics Report 69 (1972).

(16) N Levinson: A second order differential equation with singular solutions. Ann. Math. 50 (1949), 127-153.

(17) V S Manoranjan: Numerical studies of reaction-diffusion problems (this Proceedings).

(18) H P McKean: Nagumo's Equation, Advances in Mathematics 4 (1970), 209-223.

(19) R M Miura: Accurate computation of the stable solitary wave for the FitzHugh-Nagumo equations, J. Math. Biol. 13 (1982) 247-261.

(20) J Nagumo, S Arimoto and S Yoshizawa: An active pulse transmission line simulating nerve axon. Proc. I.R.E. 50 (1962) 2061-2071.

(21) S Newhouse and J Palis: Cycles and Bifurcation Theory, Asterisque 31 (1976) 43-141.

(22) J P Pauwelussen: Heteroclinic waves of the FitzHugh-Nagumo equations, Amsterdam Univ. Report TW 209/80 (1980).

(23) J Rauch and J Smoller: Qualitative theory of the FitzHugh-Nagumo equations. Advances in Mathematics 27 (1978) 12-44.

(24) J Rinzel and J B Keller: Travelling wave solutions of a nerve conduction equation. Biophys. J. 13 (1973) 1313-1337.

(25) F Rothe: Convergence to pushed fronts, Rocky Mountain J. Math. 11 (1981), 617-633.

(26) M E Schonbek: Boundary value problems for the FitzHugh-Nagumo equations, J. Differential Equations 30 (1978), 119-146.

(27) B D Sleeman: FitzHugh's nerve axon equations, J. Math. Biol. 2 (1975), 341-349.

(28) B D Sleeman, Instability of certain travelling wave solutions to the FitzHugh-Nagumo nerve axon equations, J. Math. Anal. Applics. 74 (1980), 106-119.

(29) B D Sleeman: Small amplitude periodic waves for the FitzHugh-Nagumo equations, J. Math. Biol. 14 (1982), 309-325.

(30) S Smale: Differentiable dynamical systems, Bull. Amer. Math. Soc. 73 (1967), 747-817.

(31) W C Troy: Bifurcation phenomena in FitzHugh's nerve conduction equations, J. Math. Anal. and Appl. 54 (1976), 678-690.

(32) K Uchiyama: The behaviour of solutions of some non-linear diffusion equations for large time, J. Math. Kyoto Univ. 18 (1978), 453-508.

(33) A T Winfree: Peculiarities in the impulse response to pacemaker neurons, Lectures in Appl. Math. 19 (1981), 265-279.

Department of Mathematical Sciences
University of Dundee
DUNDEE DD1 4HN

The physiology of the nerve action potential: a guide for the perplexed

N.C. SPURWAY

BASICS

In roughly the 100 years from 1840, physiologists had come to appreciate that the mode of propagation of information down long nerve fibres was essentially electrical: a brief pulse (definable now as ~1ms), negative-going as seen from outside. Originally termed the "action current", the unit signal later came, with no improvement of precision, to be called an "action potential" (A.P.) - actually, of course, a potential *change*.

The equipment available, though improving massively through the century in both speed and sensitivity of response, was based always on electrodes placed external to the nerve fibre, and so able to record only the currents set up in the surrounding medium by the nerve's activity. Nevertheless, it was possible with such equipment to show that the A.P. magnitude, though dependent upon the condition of the nerve, was independent of the strength of the initiating stimulus - provided that strength was sufficient to produce a response at all. Such responses are designated "all-or-none", and the dividing line between adequate and inadequate stimuli is the "threshold". In a given fresh nerve, threshold was a fairly constant quantity. Fatigue, and almost all other forms of biological rundown, raised the threshold - i.e. made the nerve less excitable - and reduced the size of the subsequent response. So did the very recent occurrence of a previous A.P.; in this instance the nerve is said to be, for 1-3ms afterwards absolutely, and for say 4-10ms after that relatively, "refractory".

At the end of the historical period being sketched, three other results appeared which may be said to have marked the beginning of the modern, biophysical study of nerve. One was that the velocity of propagation of the A.P. (loosely termed its "conduction" velocity) was greater in nerves of greater diameter. Another was that this same velocity also increased with the conductivity of the external medium - a result which was true even if that conductivity had been enhanced by something as abiological as a metallic bridge. The third result was that the electrical impedance of the nerve membrane, measured transversely to the fibre axis, greatly diminished during the brief period of the A.P.'s passage. The latter two of these findings were due respectively to A.L. Hodgkin and to K.C. Cole, in whose two laboratories understanding of the mechanism giving rise to the A.P. was to be revolutionised during the decade following the second world war.

INTRACELLULAR RECORDING

The first great step was Hodgkin's realization that giant fibres in squid and related creatures, newly recognised as nerves by J.Z. Young, were large enough for fine capillary tubes filled with saline to be pushed down their axes. With a wire dipped into its fluid, such a tube could serve as an *intracellular* electrode. Completing the recording circuit with a second electrode, placed in the bathing solution distant from the nerve (an "indifferent" electrode), it was possible for the first time to record the potential difference between inside and outside - the "membrane potential" (in general, E_m). At rest, this p.d. was much as had long been suspected: inside was negative with respect to outside by some 50-60mV. During an A.P., however, there occurred not merely a depolarization (which had also long been predicted) but an "overshoot" to as much as 30-40mV *positive* (Fig. 1). This overshoot was new and crucial. We must, however, discuss the rest-state first.

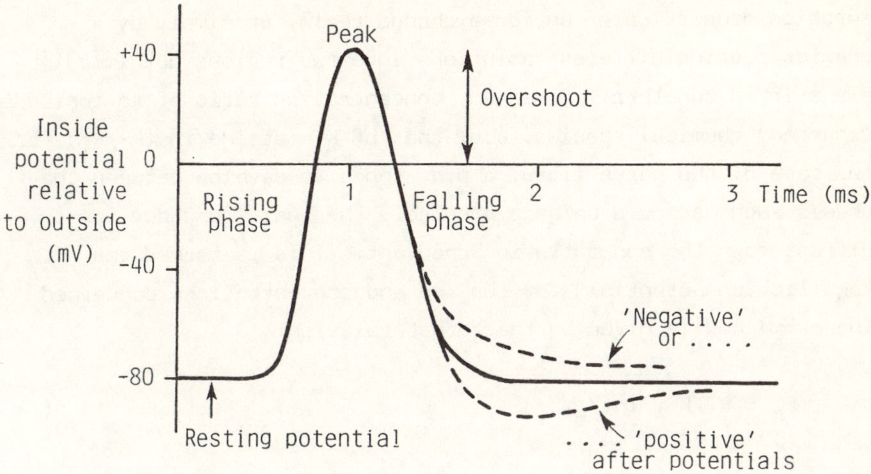

Fig.1. Representative nerve action potential, and terms used to describe various of its features. Note that resting potentials range from ~-60mV (squid) to ~-90mV (mammal).

IONS AND THE RESTING POTENTIAL

Nerves, like most cells, contain a lot more potassium (K^+ ion) than sodium (Na^+). Yet the "extracellular fluids" - outside the cells but within the animal, separated from the true exterior by the skin - contain much more sodium. Metabolically-fuelled processes - Na^+ or Na^+/K^+-exchange "pumps" - maintain this difference by driving out Na^+ and letting K^+ enter; with the first form of pump it enters passively, as a substitute, with the second by a reciprocal active uptake. Those pumps which produce net outflow of positive ions, and so a continuous outward current ("electrogenic" pumps) contribute directly to the internal negativity; at normal rates of working, however, their contribution is only a few mV.

The greater part of the resting transmembrane p.d. - "resting potential", for short - has a different cause. It does not draw directly upon the pump, but only on the ionic separation the pump produces. The mechanism would apply in any physico-chemical system which displayed the kind of separation described, even if the situation had been set up, not by a metabolic pump, but by the sorption properties of an ion-exchange resin, or simply by a chemist pouring different solutions into two regions not immediately stirred together. Given any concentration ratio of an ionised (charged) chemical species, e.g. that of K^+ outside/in (K_o/K_i) for the case of the nerve fibre, a p.d. tends to develop between the phases such as would balance precisely the ionic tendency to diffuse down the concentration gradient. This is termed the "equilibrium potential" for the ion and concentrations concerned (here E_K) and is given by the Nernst relation:

$$E_K = RT/F \cdot \ln(K_o/K_i). \tag{1}$$

(In this expression R, the gas constant, converts molar to thermodynamic units; T, absolute temperature, embodies the fact that faster-moving ions need a greater voltage to restrain them; and F, the charge per mole, expresses the reciprocal fact that ions with higher charge would be more susceptible to the electric field.)

The Nernst relation is a thermodynamic result applying to a single ionic species. All real systems are multi-ionic - the biological one particularly so. The steady-state p.d. between two real phases will be a weighted mean of the equilibrium potentials of the various ions present. The weighting factor for each ion expresses the relative ease with which it can move between the phases, and therefore carry current. Thus E_K in squid nerve is ~-60mV. E_{Na}, of course, has similar magnitude but opposite sign. However the permeability of the resting nerve membrane to Na^+ (P_{Na}) is about .01 of that to K^+ (P_K). Thinking of the two ionic concentration ratios as independent batteries, whose resultant is the transmembrane p.d., E_K operates through about 1/100th the resistance confronting E_{Na}, and so has much the dominant influence. We may conclude that, from this cause alone, the value of the overall membrane potential in a resting nerve cell (E_r) will be inside-negative by an amount only a little less than E_K. Where the direct electrogenic effect of a pump is added to this, the simple thumb-rule, $E_r \sim E_K$, will come even closer to the truth.

SODIUM THEORY OF THE ACTION POTENTIAL

The account just given, of why $E_m \sim E_K$, had in considerable measure been anticipated by Bernstein in the period 1868-1912. But Bernstein's notion of what happened during the A.P. was that the selective permeability of the membrane, for K^+ against Na^+, simply broke down, so that at the peak of the response the membrane was effectively short-circuited. What Hodgkin and colleagues perceived was that the overshoot could not be explained by breakdown but only by *switch* of permselectivity, to the transient favour of an ion for which the equilibrium potential was inside-positive. The obvious candidate was Na^+ - particularly as the high currents, required to redistribute the charge sufficiently quickly from one side of the membrane to the other, suggested an ion present in generous concentration. The basic test of this idea was published by Hodgkin & Katz (1949): they showed that the extent of overshoot and rate of voltage-change both fell when external Na^+ concentration was reduced.

So the "sodium theory of the A.P." was that, on adequate stimulation, P_{Na} increased massively (3-4 orders of magnitude). As a consequence, large numbers of Na^+ ions would enter the nerve in a short time, driven by both concentration and electrical gradients, until the inside potential had moved positive relative to the outside and brought $E_m \sim E_{Na}$. This would be a new, near-equilibrium condition, in which net inflow of Na^+ was reduced to a trickle - not actually zero because P_K was not zero, and therefore E_m would settle at a value slightly $< E_{Na}$. This situation could be maintained indefinitely, with inward Na^+ flow (I_{Na}) = outward K^+ flow (I_K) - i.e. with no net current flowing across the membrane and so no further voltage change. In fact, however (the theory continued) even before the near-equilibrium point had been reached, P_{Na} started to change back again towards its resting value; and, as a consequence of this alone, E_m would fall back also.

The account has not finished here, but we can make two comments at this point. First, it is now easy to see why the A.P. is observed as a negative swing of outside voltage simultaneous with a positive swing of inside voltage: positive charge, in the form of Na^+ ions, leaves the external fluids to enter the nerve. Second, the nature of threshold and the all-or-none response can be illuminated if the process by which P_{Na} rises is considered more fully. The fundamental phenomenon of excitability is that P_{Na} increases with depolarization. Subthreshold depolarization has such an effect in a small way; but outward I_K, flowing through the relatively-permeable resting K^+ channel, more than counters the electrical consequence of the slightly enhanced Na^+ entry. K^+ outflow thus restores the resting potential and turns the P_{Na} increase off again. Suprathreshold depolarization, however, raises P_{Na} to a level at which inward I_{Na} exceeds the greatest outward I_K attainable through the normal conductance; so further depolarization results. But this is a positive feedback: the further depolarization further increases P_{Na} ... and so on, regeneratively, till the maximum P_{Na} and maximum overshoot have both been reached. All responses will reach this maximum, those to large initial depolarizations simply doing so rather faster than those to stimuli which were just supra-threshold. Threshold itself is a potential at which $I_{Na}=I_K$. Unlike the other, more-theoretical potential having this property (the maximum possible overshoot, mentioned above), this one is unstable: it may linger for some ms before random fluctuations turn it, with equal probability, either upward towards a regenerative A.P. or downward again to the resting state.

POTASSIUM CURRENTS AND THE MEMBRANE CAPACITANCE

Early in the previous section we recognized, in passing, that a high current must flow if membrane potential is to be changed fast. The electrical formalization of this idea is of course embodied in the concept of "capacitance", which is measured by the quantity of charge that must be moved from one conductor to another for unit p.d. to be developed between them. To apply this to nerve is intuitively easy, for the membrane (a fatty layer, normally of substantial electrical resistance) separates two volumes of salty fluid (good conductors): the analogy with a parallel-plate condenser is obvious. Because the membrane is extremely thin

(~7nm) the electrostatic interaction between one conductor and the other is considerable. Thus a large charge must be transferred to develop unit p.d. between them - i.e. the capacitance per unit area of membrane (C_m) is high. A typical, round value is $1\mu F/cm^2$.

The time constant for membrane voltage-change is then given by C_mR, where R is the resistance through which current is flowing (time in secs, C in farads, R in ohms). Considering only the resistance of unit area of the membrane itself (R_m), we can quickly calculate the "membrane time constant" (C_mR_m) for representative circumstances. This provides an upper bound for dE_m/dt in each condition. (Solution resistances, in series with R_m, extend the overall time constant, but not often sufficiently to destroy the utility of the present, order-of-magnitude calculation.) R_m is $\sim 25\Omega\text{-cm}^2$ when P_{Na} is at its peak, but $\sim 1000\Omega\text{-cm}^2$ in an unexcited membrane. Taking $5C_mR_m$ as the time in which voltage response to a step change of e.m.f. is effectively complete, we see that the rising phase of the A.P. is likely to take only ~0.1ms longer than the rate of P_{Na} increase ... whatever that turns out to be. By contrast, even in the unlikely event that, after the peak of the A.P., P_{Na} cut back to near zero virtually instantaneously, E_m could not take less than about 5ms to fall back to resting value in response. Experimental records (e.g. Fig. 1) plainly show that it falls faster.

The explanation offered, as early as the first statement of the "sodium theory", was that, at the same time as P_{Na} was falling, P_K rose. This would allow K^+ to move outwards, and consequently restore E_m to $\sim E_r$, faster than we have been assuming. In formal terms, one would say that:

<u>i</u> The equilibrium value, toward which E_m is moving, would more rapidly return to $\sim E_K$ - recall that this equilibrium value is a mean of E_{Na} and E_K, weighted inversely to the individual resistances through which they act.

<u>ii</u> Overall R_m, and hence the time constant governing the approach of E_m to the assymptote just indicated, would be lower than we calculated in the previous paragraph.

To summarize, the theory at this point is that an adequate depolarizing pulse is one which sufficiently increases P_{Na} for further depolarization to ensue, and P_{Na} to run away to its maximum; but that an inbuilt time-dependence limits the duration of this response, which is promptly succeeded by an increase of P_K, accelerating restoration of the *status quo*.

PROPAGATION OF THE IMPULSE

We are poised now to ask for kinetic details of the postulated Na^+ and K^+ conductance changes. Later exposition will, however, be facilitated if we first look at the process whereby an A.P. propagates along an "axon" (long nerve fibre), for so far we have concentrated all detailed discussion on the time- and voltage-dependent events at any one place. Much of Hodgkin's earliest work (1939 *et prec.*) was on propagation; and he dealt with the complexities of current-flow, in effectively infinite volume conductors, by speaking in terms rather reminiscent of Faraday's "lines of force" - he spoke of "local (electrical) circuits". From the transiently-positive domain of intracellular fluid, just deep to an active site on the membrane, current would flow along the core of the axon, out through the membrane at varying distances ahead, and back through the extracellular fluids towards the region of negative potential immediately outside the active site. Such a current, considered at the point where its "circuit" carried it radially outward through the fairly high resistance of the unexcited membrane, would raise the internal potential and lower the external - i.e. depolarize the membrane. Where the depolarization was suprathreshold, excitation would occur. Thus the A.P. would propagate.

Notice, in passing, two points about propagation-direction. First, that the directionality of impulse traffic *in vivo* arises simply from the fact that a nerve is normally stimulated at one end; if, in an experiment or mishap, it is stimulated in the middle, propagation occurs in both directions. What stops this situation from causing chaos in the nervous system is that the nerve-nerve junctions - "synapses" - permit communication only one way. Thus any contra-normal impulse flow does not get beyond the axon in which it started. The second point arises out of these comments, for it must now appear that, from any location excited during normal, forward propagation, re-excitation backwards could occur. (One may have met similar problems in seeking physical understanding of, say, light wave-trains.) In the nerve case, the explanation is built into our subject: the region just behind the peak of any A.P. is refractory.

Plainly, the greater the 'reach' (informal term), ahead of an active region, of the zone of suprathreshold depolarization, the greater will be the velocity of propagation of the signal down the fibre. This forward reach of adequate depolarization is an inverse function of the nerve's axial resistance (itself $\alpha[\text{diameter}]^{-\frac{1}{2}}$) and also of external resistance. Hence the increase of velocity with fibre diameter in biological circumstances, and with the conductivity of the external medium in experimental ones.

CABLE THEORY

Formal treatments of current spread and conduction velocity (e.g. those of Hodgkin & Rushton, 1946 and Rushton, 1951) invoke the evident analogy between an axon and a submarine cable. In a simple, cylindrical cable of radius *a* (Fig. 2), bathed in a bulk

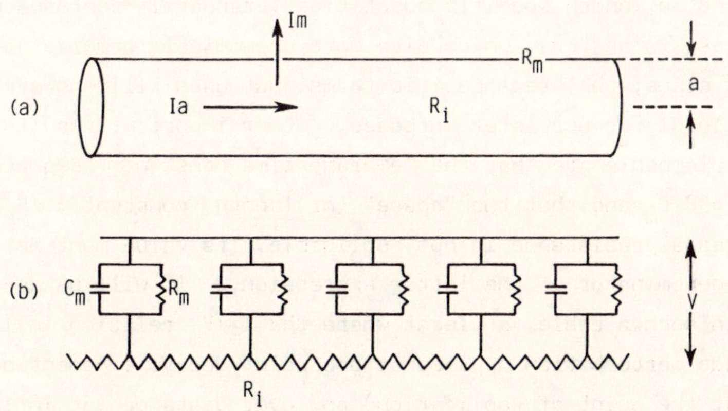

Fig.2 (a). The axon treated as a simple cable. (b) Its electrical equivalent circuit, where external resistance is negligible. V is measured as displacement from E_r.

medium of negligible resistance, let I_a and I_m represent respectively the axial current and that passing radially through unit area of membrane; R_i is the intracellular resistivity (resistance of a unit cube of intracellular fluid) and R_m the resistance of unit membrane area; and V is a signal voltage. (A symbol unrelated to E_m is chosen for the latter because, in considering experimental work associated with the theory, one is concerned with injected current pulses as well as with transient changes in the resultant of the ionic e.m.fs.) From Ohm's law:

$$dV/dx = -I_a R_i/\pi a^2$$

while from Kirchoff's:

$$dI_a/dx = -2\pi a I_m \ .$$

Differentiating the first and substituting from the second we conclude:

$$I_m = (a/2R_i)d^2V/dx^2 \ . \tag{2}$$

This is termed the "cable equation". It may alternatively be expressed in length-specific quantities (standardly represented by lower-case *i*s and *r*s), which give more economical formulae at the present stage; the area-specific terms just used will, however, be more helpful for our later purposes. Other important results for the simple cable are that the membrane time constant (as quoted above) = $R_m C_m$ and that the "space" (or length) constant = $\sqrt{R_m a/2R_i}$. (If external resistance is not negligible, its value must be added to the denominator of the latter expression.) It will be apparent that, in such a cable, at least where the I_m/V_m relation is linear, a voltage perturbation applied at one point decays exponentially in time (at the point of application) and over distance (at infinite time). Note, however, that the approach to the steady state at points other than that of application must be characterized in terms of error functions.

THE VOLTAGE CLAMP

About 1950, with the sodium theory stated qualitatively and the cable characteristics of the axon known quantitatively, Hodgkin and coworkers naturally wished to obtain comparably quantitative data describing the kinetics of conductance change. The key to so doing would be a means of isolating the experimental variables. Equipment capable of this had just been developed in Cole's laboratory; it was to become known as the "voltage clamp".

Two electrodes had to be inserted along the core of the axon. One was metallic, and uninsulated over the last several mms of its length. This made R_i negligible, and consequently eliminated the length-variable except at the ends of the electrode - and the regions of the axon there were excluded from study by guard-rings. The studied length of membrane now performed as a synchronous unit and was described, in respect of this feature alone, as "space-clamped". To achieve voltage-clamping, the space-clamping electrode was made to carry current to the preparation from an external source. The rest of the equipment functioned to adjust the value of this current so that, after an initially-imposed voltage step, the transmembrane voltage in the space-clamped region did not alter further - i.e. it was "voltage clamped" as well. The second electrode provided the input to this additional circuitry: it was a voltage-sensing electrode of the sort already familiar, inserted to about midway along the clamped region, and reading potential with respect to a remote, external electrode in just the way described above. However, disparity between this reading and the command potential set by the experimenter was here fed back, to adjust the output of the current source. In control-theory terms, any mismatch between set and measured voltage constituted the error signal in a negative feedback loop. Although the electronic properties required of the current source are more readily achieved nowadays, with "operational amplifiers" based on solid-state devices, a close approach to voltage control was obtained using thermionics in 1949-52. (Cole had been trained as an electrical engineer, and both Hodgkin and Huxley had spent their war-service developing radar.)

Voltage constancy implies zero net current across the membrane. Once the "capacitative transient" was over (this being the surge of current required to impose the voltage step upon the membrane capacitance) "zero net current" meant that the total ionic current through the membrane at any instant (the "intrinsic current") must be equal in magnitude and opposite in sign to the output current from the electronic source ("extrinsic current"). Plots of the variations of extrinsic current with time constituted the experimental records. Thus, while the voltage-dependence of intrinsic current could be investigated by collating responses to many different command voltages, during each individual record dI_m/dV, dI_m/dx, dV/dx and dV/dt were all zero: only dI_m/dt was not, and could be unambiguously studied.

CLAMP RESULTS

The results cited in this section will be taken from the series of papers: Hodgkin, Huxley & Katz (1952); Hodgkin & Huxley (1952 a, b,c). We will consider only an axon held. for many seconds prior to the main observation, at an internal potential typical of the ordinary resting state (say, -60mV) and only the ionic currents - i.e. the currents flowing after completion of the capacitative transients. Currents obtained when the axon was bathed in normal, Na^+-rich, extracellular fluid will be compared with those in a medium containing no Na^+. (So as not to alter solution parameters such as osmotic pressure, Na^+ was never simply omitted: an organic cation, choline$^+$, was substituted. This ion was justifiably considered too large to pass through channels selective for any such monatomic cation as Na^+.)

If the command step drove E_m to +40 or more mV, all current-flow was outward; the only difference (Fig. 3) between the two solutions was that the response showed slightly more delay with Na^+ present. Less massive positive steps, imposed in Na^+ media, elicited biphasic flow: first in, then out. Peak inward flow was obtained with a depolarizing step of ~40-60mV, and lasted ~1ms (though longer at 0°C than 20°C). Outward flow then followed, and lasted till the command depolarization was turned off. In the

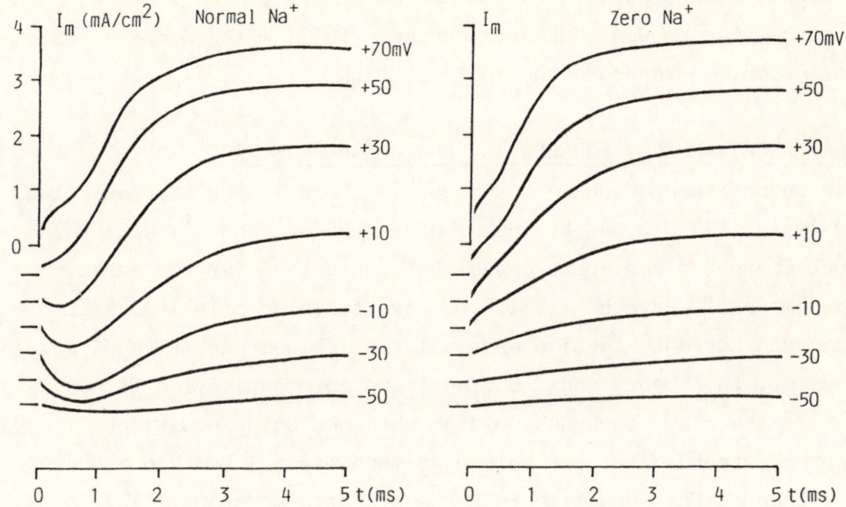

Fig.3. Membrane currents (upward deflections = outward) in squid axon, $E_r \simeq -60$mV, voltage clamped to internal potentials indicated, right. Freehand sketches for, say, 10°C, based on results of Hodgkin and colleagues (1952). Transient capacitative currents omitted.

absence of Na^+, however, no size of depolarizing step elicited any phase of inward flow.

The interpretation accepted was that, in the first ms, the net ionic battery (which, according to the sodium theory, should in this phase be heavily weighted toward E_{Na}) had an e.m.f. ~+40mV. Command voltages more positive than this would always drive cations (and therefore current) outward, but less positive imposed voltages would allow cations to flow inward during this period. Obviously, if the theory is right, the actual inflow is of Na^+: and this is supported strongly by the clamp records showing that inward current is eliminated by the choline$^+$ substitution. By contrast, the late current ought to consist in K^+ outflow. Fig. 3 indeed shows that this current does not depend on $[Na^+]_o$, but direct test required lowering of *intracellular* K^+ concentration to zero. This was technically much harder than modifying extracellular solutions: the means to do it would not be found till 1960 (see below). Meanwhile, late, Na^+-independent currents were provisionally accepted as more or less totally attributable to K^+.

One other result must be noted. Though we have seen from Fig. 3 that both Na^+ and K^+ currents increase with a delay when the axon is depolarized, early and later terminations of the imposed voltage step showed that both currents fall, *with no appreciable inflexion*, in response to repolarization.

THE HODGKIN-HUXLEY FORMULATION OF CHANNEL KINETICS

The objective of Hodgkin & Huxley, in their famous theoretical analysis (1952d), was to test whether the Na^+ and K^+ currents, just characterized, were capable of accounting for the various phenomena of nerve excitability. Tractable descriptions of the observed current kinetics were all that this strictly required. However, the theory would obviously be more satisfying if adequate fit to the clamp records could be obtained using equations appropriate to plausible molecular mechanisms. Hodgkin & Huxley had thought for some time in terms of intra-membrane, flux-controlling particles, which were themselves charged and therefore voltage-sensitive. By 1952 they had fined this model down to that of "gating" particles - elements which did not themselves carry ionic current, but acted permissively. (Among the reasons for regarding the gating process as being distinct from ion-carriage itself was that the rates at which currents rose and fell were steeply dependent on temperature but the peak currents reached appeared not to be.) In the formalization, each gating particle was envisaged as switching binarily, between one position permitting ions to cross the membrane and another position preventing passage. First-order kinetics were assumed. Thus the equations used were of the form:

$$dy/dt = \alpha(1 - y) - \beta y \qquad (3)$$

where y (dimensionless) varies between 0 and 1, and could represent the fraction of the gating particle in one of its two positions. α and β are voltage-dependent rate constants, describing the rates of entering and leaving the y-location. The original authors emphasized, however, that their quantitative test of the sodium theory did not hinge upon a particular molecular account of the gating process, which was offered only to explain the form of

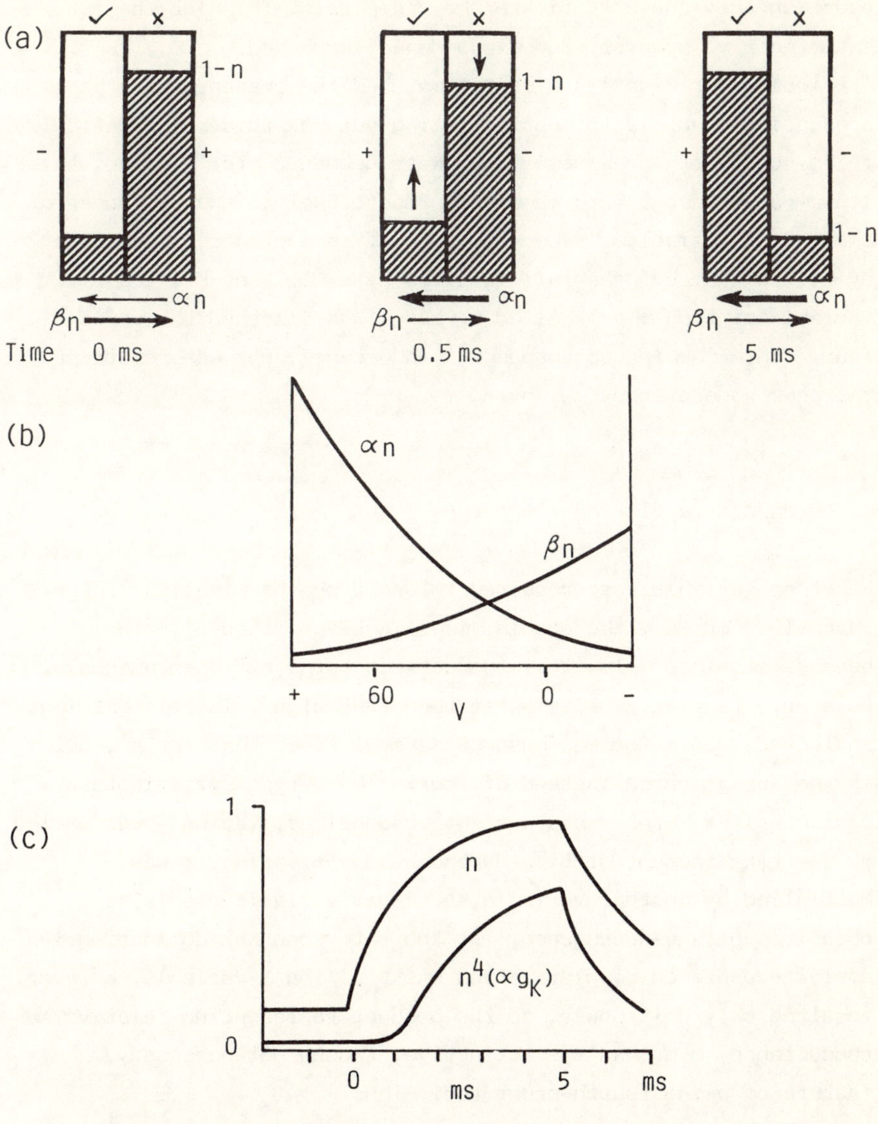

Fig.4. Processes assumed by Hodgkin & Huxley to determine K^+ conductance in squid nerve. (a) Response of n-particles to sudden depolarization. Initially, fraction n in conductance-permitting (✓) location is small, since $\beta_n > \alpha_n$. Following depolarization, $\alpha_n > \beta_n$, and n rises exponentially at the expense of the non-conducting location (×). (b) Variation of rate-coefficients, α_n & β_n, with V (where V = 0 is resting condition). (c) Responses of n and g_K to depolarization (by ~ 60mV) lasting 5ms. Sketch-curves only. Based on Jack, Noble & Tsien (1975), Fig.8.9 (part).

equation they chose to fit to the clamp data. Finding the fit satisfactory, however, Hodgkin & Huxley proceded.

First order kinetics directly matched the responses of the conducting channels to repolarization but not to depolarization: third- or fourth-order equations were needed there. To meet this, ionic conductances were supposed proportional to the *a*th power of the *y*-type variables. Then the rise of conductance is described by $(1 - e^{-t})^a$, but the fall is given by e^{-at}. For K^+, the variable of type *y* was called *n*, and *a* was allotted the value four. Then the instantaneous K^+ conductance, g_K, was related to the peak conductance, $\bar{g}_K$, by:

$$g_K = \bar{g}_K n^4$$
$$dn/dt = \alpha_n(1 - n) - \beta_n n \ .$$

No other variables are required to describe the kinetics of the K^+ channel. For Na^+, the problem is slightly more complicated, because at times >~1ms this conductance turns off spontaneously. However, its rise, while faster than that of g_K, is somewhat less inflected. So α for Na^+ must be several times that for K^+, but *a* may be set at three instead of four. The *y*-type variable thus characterised - the measure of Na^+-channel activation - was called *m*. To take account of time-dependent *inactivation*, *m* was multiplied by another variable, *h*. This variable displayed opposite voltage-dependence, so the rate constant dominating its early response to depolarization was β_h. The *h*-variable, however, required only unit power, so the product representing relative Na^+ conductance, m^3h, had the attractive, though not necessary, feature of being fourth-order too. Thus

$$g_{Na} = \bar{g}_{Na} m^3 h$$

and the kinetic equations for *m* and *h* have the familiar form but with $\alpha_m > \beta_h > \alpha_n$ for depolarizations (Figs. 4, 5).

Returning, with all caveats, to the migrating-particle model, *m* and *n* could be thought of as refering to two kinds of negatively-charged particle, each held at the outer surface of the membrane

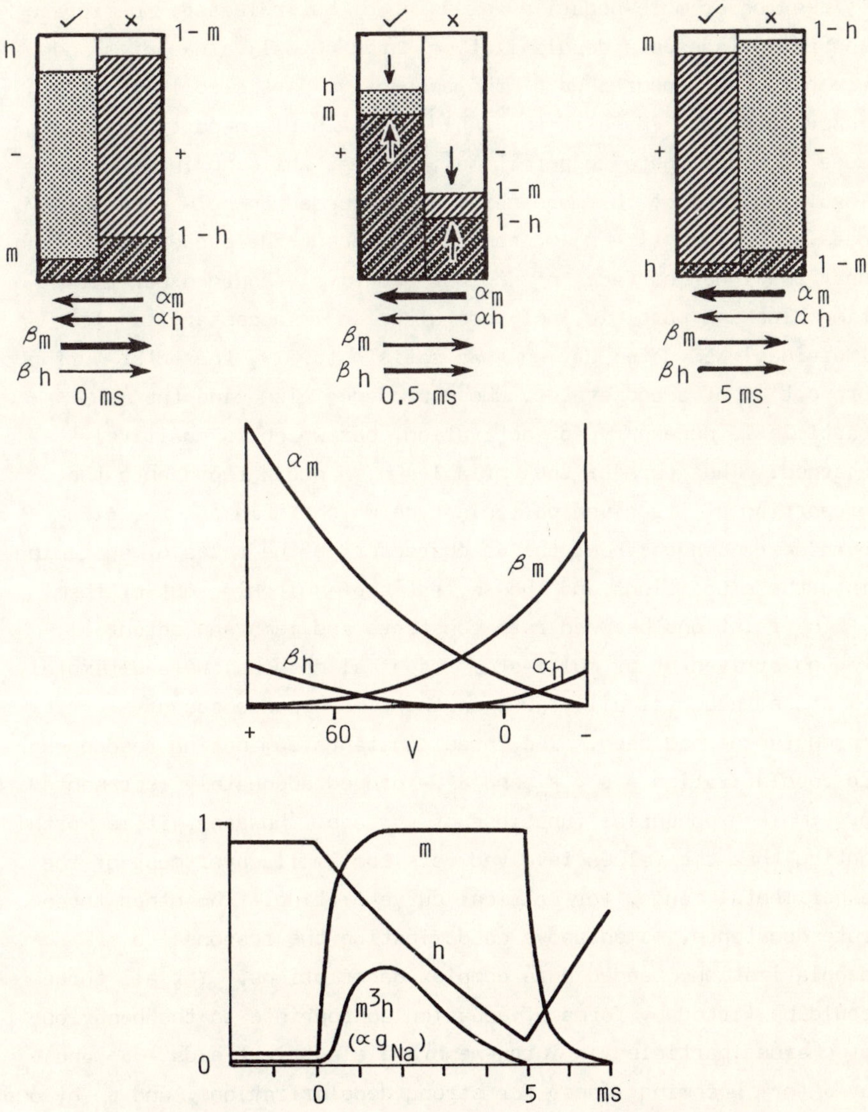

Fig.5. Processes assumed to determine Na^+ conductance in squid nerve. (a) Responses of m (activating) and h (inactivating) particles to sudden depolarization. Initially, m small, h large since $\beta_m > \alpha_m$, $\alpha_h > \beta_h$. On depolarisation, $\alpha_m > \beta_m$; also $\beta_h > \alpha_h$ but $< \alpha_m$. So m rises markedly before h falls far. As in Fig.4, ✓ = conductance-permitting, × = non-conducting location. (b) Variations of rate-coefficients with V. (c) Responses of m, h and g_{Na} to depolarization (by ~ 60mV) lasting 5ms. Based on Jack, Noble & Tsien (1975) Fig.8.10 (part).

by resting or more-negative values of E_m but released, to migrate inward, by adequate depolarization: the actual values of *m* and *n* are then the proportions of the respective particles lying at the inner surface. Three of the rapidly-switchable *m*-particles must come together there to permit Na^+ passage, while an interaction involving four of the more tardy *n*s is needed to hold open the K^+ gate. The term (1 - h) might then represent the proportion, lying near the inner surface, of another negatively-charged but this time blocking particle, only one of which is necessary to close an individual Na^+ gate; or, just as satisfactorily, the outer-surface proportion of a cooperator, whose presence alongside the *m* particle is necessary for activation, but which is positively charged. Thus each of the variables *m*, *h* and *n* represents the proportion of its given particle-type which is in *the position permitting conductance*; the αs characterize the rates of switching into these positions and the βs, rates of switching out of them.

The relations between rate constants and membrane potential, though arrived at by multi-step graphical methods, were ultimately expressed algebraically, just as the time-dependences of conductances had been. The three constants dominating responses to repolarization - β_n, β_m and α_h - proved adequately represented by simple exponential functions of voltage. However, it is worth noting that the values involved were too small, over most of the experimental range, for critical curve-fitting. The other three rate constants, named above as dominating the response to depolarization, needed more complex descriptions. Yet all three could be fitted by forms of equation appropriate to the behaviour of charged particles in intra-membrane electric fields - α_n and α_m by a form becoming linear for strong depolarizations, and β_h by one allowing for saturation in that range.

HODGKIN-HUXLEY ACCOUNT OF THE ACTION POTENTIAL

With the instantaneous conductances, g_K, g_{Na}, defined as products of peak conductances and the appropriate powers of *n*, *m*, *h*, and the time- and voltage-dependences of these terms embodied in the expressions for α_n, β_n, &c, Hodgkin & Huxley had expressed the sodium theory quantitatively. What remained was to test it, by seeing if it could account quantitatively for the phenomena of nerve excitation - in particular, the A.P..

If they were right, the complete expression for the membrane current density (I_m) at any moment was:

$$I_m = C_m dV/dt + (V - V_K)\bar{g}_K n^4 + (V - V_{Na})\bar{g}_{Na} m^3 h + (V - V_L)\bar{g}_L \quad (4).$$

Here all symbols have been previously defined except V_K, V_{Na} and those in *L*. V_K and V_{Na} are the differences between the equilibrium potentials for the respective ions and resting potential, i.e. $E_K - E_r$, $E_{Na} - E_r$. Terms in *L* take account of the fact that any other ionic permeabilities the membrane possesses will constitute a "leak" conductance, acting in parallel with those to Na^+ and K^+. Note that the theory has at no point excluded these, only excluded them from *changing*; and we might add that the experimental results, for membrane resistance and for the extent of overshoot, indicate that they cannot be large. Thus g_L, though small, is not zero, but in a given solution it is constant. The main ion likely to be involved, being both small in size and present in bulk, was Cl^-; so E_L, the net equilibrium potential of the leak ions, must be close to E_{Cl}.

In all three ionic terms, one other important empirical result is embodied: the current in the channel is represented as the simple product of the channel conductance and the net driving force on the ion(s) concerned - i.e. as obeying Ohm's law. The validity of this had been demonstrated by Hodgkin & Huxley (1952b) when they showed that the "instantaneous current/voltage relation", namely that measured before the *g* values had begun to adapt to a new *V*, was linear in each channel. The experiments demonstrating this used two successive steps of command voltage. An alternative situation, in which non-ohmic (e.g. rectifying) instantaneous behaviour would again show up, would be where results were compared from different solutions, such that the equilibrium potentials were not the same.

The first condition for which equation (4) was solved was that for a length of axon which was space-, but of course not voltage-clamped. The result, called a "*membrane A.P.*", was the response given by the whole studied length of membrane acting in synchrony. A short shock is applied to the long metal wire in the axon core. After it, ionic currents merely charge and discharge C_m: so I_m is zero for $t>0$, and the boundary condition is the initial displacement of V. Solutions, worked out initially with a hand-turned calculating machine, agreed very well with the behaviour of a real nerve.

The biologically important A.P. is, however, that which migrates from end to end of an unclamped nerve. To test for the theory's ability to account for this "*propagating A.P.*", two equations for I_m must be solved simultaneously. These are (4), giving I_m in terms of ionic currents and capacitative voltage-change at a particular point of the membrane, and (2), relating I_m to the spatial gradient of voltage along a cable. Simultaneous solution cannot be achieved in general, owing to the presence of differentials in both x and t, but it can for the specific case of a wave-like disturbance moving along the cable at constant velocity - which is what the propagating A.P. is. In this situation, x may be replaced by $-\theta t$ (where θ is the velocity), and (2) becomes:

$$I_m = (a/2R_i\theta^2)d^2V/dt^2 \qquad (5).$$

When the two expressions for I_m are now brought together, θ is an unknown constant. The procedure used by the original authors was to guess a value and start a trial solution. V goes to $\pm\infty$ according to whether θ has been chosen too high or too low. The correct value is that which brings V (the *displacement* of membrane potential) back to zero at the end of the computational run. Agreement with experiment was achieved within about 15%. Considering the unavoidable approximations in the earlier curve-fitting, this was probably as good as could have been hoped. The shapes of both membrane and propagating A.P.s also fitted very acceptably.

The final outcome of a majestic combination of insight and effort was the widely-quoted summary plot, Fig. 6.

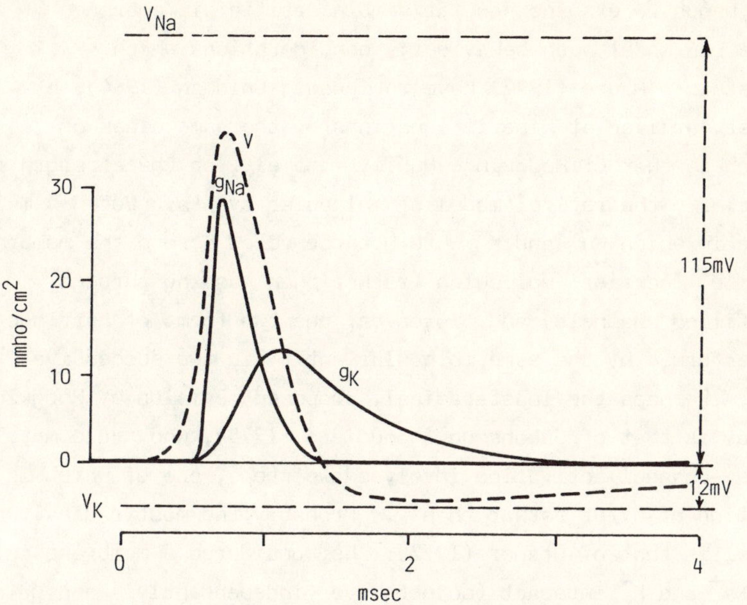

Fig.6. Theoretical solution for propagated action potential and conductances to Na^+ and K^+ at 18.5°C. (From Hodgkin & Huxley, 1952d).

OTHER MATHEMATICAL TREATMENTS

Cooley & Dodge (1966) re-analysed the original Hodgkin-Huxley equations on a digital computer, with some improvements of numerical method made feasible by the automation. See also Palti's chapters in Adelman (1971). Fitzhugh explored a generalized and somewhat simplified series of equations in a number of papers (reviewed 1969). This and other simplifications have also interested Noble and colleagues: see Noble (1973), Jack et al. (1975), and prior references in both. These authors have also made considerable use of dimensional analysis: see particularly, Fitzhugh (1973), Hunter et al. (1975).

Hoyt (1963) achieved superior fit to the original experimental plots with different kinetic equations from those of Hodgkin & Huxley, notably one which described both activation and inactivation of the Na^+ channel in terms of a single, second-order parameter. Arndt & Roper (1974) also sought equations giving statistically-optimized fit to clamp data, and found that the activation terms described the conductance best when raised to the second power, rather than the third and fourth. They have since

(1975) begun to examine how far simple artificial membrane systems can model such behaviour: considerations which Finkelstein & Mauro (1977) have reviewed. Goldman (1964), by contrast, arrived at kinetic equations by the same class of approach as that of Hodgkin & Huxley - namely, on the strength of an *a priori*, theoretical model of molecular events. But his model was one in which Na^+ and K^+ were both conveyed across the membrane by mobile, 'carrier' molecules (rather than passing through water-filled channels) and, moreover, the two forms of carriage were performed by the same molecular entity in two successive states. Perhaps the least-radical, proposed revision by Hodgkin & Huxley is that of Jakobsson & Scudiero (1975), who argue merely that Na^+-channel activation involves two steps, one of which is a function of dV/dt rather than V. Probably the most radical proposal is that of Offner (1972): he considered a membrane in which Na^+ and K^+ interact (do not move 'independently') and do not flow proportionately to the thermodynamic gradient.

All these approaches are experiment-related and all of them (though some in literal and others in simplified ways) either embody or suggest mechanistic models. These are features which appeal to physiologists. Very general formulations such as catastrophe theory (Zeeman, 1972) seem less likely to indicate particular mechanisms for which test-experiments may be designed; they therefore appeal less. On the other hand they are not, of course, physically false. What is physically false is the assumption that there is measurable electromagnetic inductance in the axon (Lieberstein, 1973). It is true that the delay, between application of a voltage step and maximum outward current flow, represents a phase lag and therefore has one qualitative feature of an inductance (c.f. Hodgkin & Huxley, 1952, p. 540; Jack et al. 1975, p. 21). This is a feature which Cole has studied extensively (see 1968). However, dI_m/dt is not $\propto V$, so even the formal similarity is non-quantitative. Equations including the symbol L are not, therefore considered generally helpful by experimentalists.

One lower-level source of confusion should also be mentioned: sign conventions! Into the mid-1950s Hodgkin & Huxley wrote with the convention that increased membrane polarization was represented positive - i.e. values of E_m were expressed as external voltage minus internal voltage. More recent work uses the opposite convention, internal minus external, which directly represents the negative-going voltage step displayed by an electrode entering a nerve or other cell from the bathing solution. Membrane currents must also be treated consistently and appropriately; and the opposite signs of intrinsic and extrinsic currents should always be borne in mind. One has even seen opposite conventions used for I_m in equations (2) and (4), which were then combined with inevitably meaningless results.

FURTHER EXPERIMENTS

The first check performed by Hodgkin and coworkers, that their analysis was along the right lines, was that exchanges of radio-isotopic Na and K were enhanced appropriately by nerve activation. Results were summarized by Hodgkin (1964). An exciting development, a decade after the classic Hodgkin/Huxley work, was the discovery that artificial solutions could be perfused down the inside of an axon, and thus, i.a., the identification of the late outward current as being due to K^+ could be directly checked (Baker, Hodgkin & Shaw, 1962). This "perfused axon" preparation is the subject of continuing work in several laboratories. Improvements in voltage clamping, to increase current output and diminish series resistances in electrodes and solutions - both of which make most difference to control when membrane conductance is at its peak - were a theme in Cole's laboratory (see 1968). Anyone now seeking best-fit formalizations of clamp data should take notice of these results.

An entirely new approach, pioneered by Verween, was the analysis of membrane noise, and its attribution to the fluctuating electrical behaviour of unstable ionic channels. Originally studied in large areas of membrane, noise is now usually investigated in the sub-micron area which can be sucked against a micro-electrode tip and is likely to contain only a few ionic channels. Noise analysis is reviewed, with thorough groundwork, by Neher & Stevens (1977).

Direct experimental tests, of some of the respects in which later mathematical analyses had differed fundamentally from that of Hodgkin & Huxley, have also been made. Early and late currents are affected independently by other ions, and by pharmacological agents (see Junge, 1981). Sodium channel inactivation can be pharmacologically eliminated without preventing activation (Armstrong et al., 1973). These last two classes of finding are hard to square respectively with Goldman's and with Hoyt's treatments, described in the previous section, and one would have thought also with Offner's. However, Cohen & Landowne (1974) find that the temperature-dependence of the ion fluxes occurring in an A.P. is much closer to that predictable by Offner's treatment than by that of Hodgkin & Huxley.

The fundamental idea of the charged gating particle has itself been investigated. Charged particles migrating through the membrane on depolarization but moving little, if at all, the other way on hyperpolarization (the driving of the internal potential more negative than E_r) should make the capacitative transients asymmetric when the effects of voltage-steps in opposite directions are compared. Such asymmetries are indeed observed though the most recent analyses suggest that only a part of them, identifiable by its property of being inactivated by a long, prior, depolarizing pulse, represents gating currents (Armstrong, 1981; Keynes et al., 1982). The respective molecular properties of gating and ion-selective ("filter") systems have even been probed (Hille, 1978). The matters touched in the last two paragraphs are reviewed more thoroughly by Miller (this volume).

ACTION POTENTIALS IN OTHER TISSUES

The faster-conducting nerves of vertebrate animals are wrapped in repeating collars of a fatty insulator called "myelin"; the excitable membrane is exposed to external solution only at the short gaps in the insulation, called "Nodes of Ranvier". Excitation in such a nerve leaps from node to node ("saltatory conduction"), and achieves greater speed thereby. The nodal membrane, however, differs from that of squid nerve in showing non-ohmic instantaneous I_m/V_m relations, different power relations for m and n, and a K^+ conductance which either inactivates with time or, in the specific case of mammalian nerve, is not activatable at all. (This last point nicely illustrates our contention that elevation of g_K is necessary in unmyelinated nerve because C_mR_m would otherwise be unacceptably long: by reason of its thickness, myelin greatly reduces C_m.)

Muscle cells (fibres) also conduct A.P.s, to carry excitation along their lengths. The vertebrate skeletal-muscle A.P. is quite like that of nerve, but many invertebrates carry charge inward, not with Na^+ but with calcium (Ca^{++}) ions. Mixed inward currents, carried partly by Na^+, partly by Ca^{++}, are common in smooth muscles and other tissues. The general rule seems to be that Ca^{++} is evolutionarily the most primitive inward-current ion, and continues to be used where it serves an internal function (such as initiating muscle contraction) on arrival; but that Na^+, being available in much greater concentrations, has been substituted where rapid conduction over long distances has had evolutionary advantages. The most complex A.P. of all, however, is that of the mammalian heart. Here all three ions take their turn - first Na^+, then Ca^{++}, finally K^+ - with different thresholds, different functions, and (in the case of K^+) several channels for one ion.

References to many of these other tissues can be found in Jack et al. (1975) and Junge (1981). On particular tissues, however, see Adrian & Marshall (1976), Beeler & Reuter (1977), Chiu et al. (1979).

FURTHER READING

Hodgkin himself has written a lucid, elementary account of the nerve A.P. (1964). Also superbly lucid, and of rather wider coverage, is Katz (1966). A modern, succinct, and very useful account of many excitable tissues is that of Junge (1981) to which several references have already been made. Adelman (1971) provides source-material relevant to many of the later sections of the present article. A work of greater formality than any of the foregoing, which no mathematical reader should be without, is another which has already been cited several times: Jack et al. (1975). On many matters beyond those noted above, Cole (1968), though difficult reading, will also be most valuable.

Of survey articles, the classic one is Hodgkin's own (1958), while Huxley (1959) added, *i.a.*, an elegant dimensional analysis. More recent, general accounts, broadly accepting the Hodgkin/ Huxley approach, are those of Noble (1973), which emphasizes cable properties, and Hille (1977), which gives more attention to ionic mechanisms. Somewhat in contrast, though equally lucid, is Jack (1976) who emphasizes difficulties in the traditional approach, to which the present article, and on the whole the above reviews also are devoted.

Finally, however, no scientist who has taken the trouble thoroughly to understand the Hodgkin-Huxley equations could fail to enjoy, and be inspired by Hodgkin's account of the work which produced them (1977).

This article has been written, in an attempt to be of assistance, by a physiologist who teaches the subject regularly, but whose own research fields lie elsewhere. Any discrepancy between the above account and that of any of the authors cited must unhesitatingly be read in favour of the other author.

BIBLIOGRAPHY

ADELMAN, W.J. (1971) Ed.. '*Biophysics and physiology of excitable membranes*.' van Nostrand Reinhold Co..

ADRIAN, R.H. & MARSHALL, M.W. (1976). 'Action potentials reconstructed in normal and myotonic muscle fibres.' *J. Physiol*. 258, 125-143.

ARMSTRONG, C.M. (1981). 'Sodium channels and gating currents.' *Physiol. Rev*. 61, 644-683.

ARMSTRONG, C.M., BEZANILLA, F. & ROJAS, E. (1973). 'Destruction of sodium conductance inactivation in squid giant axons perfused with pronase.' *J. Gen. Physiol.* 62, 375-391.

ARNDT, R.A. & ROPER, L.D. (1974). 'A new phenomenology for squid axon voltage-clamp currents.' *J. Theoret. Biol.* 48, 373-387.

Ibid. (1975). 'Theory of excitable membranes. I. A simple model for a three-state artificial membrane.' *J. Theoret. Biol.* 54, 249-283.

BAKER, P.F., HODGKIN, A.L. & SHAW, T.I. (1962). 'Replacement of the axoplasm of giant nerve fibres with artificial solutions.' *J. Physiol.* 164, 330-354.

BEELER, G.W. & REUTER, H. (1977). 'Reconstruction of the action potential of ventricular myocardial fibres.' *J. Physiol.* 268, 177-210.

CHIU, S.Y., RITCHIE, J.M., ROGART, R.B. & STAGG, D. (1979). 'A quantitative description of membrane currents in rabbit myelinated nerve.' *J. Physiol.* 292, 149-166.

COHEN, L.B. & LANDOWNE, D. (1974). 'The temperature dependence of the movement of sodium ions associated with nerve impulses.' *J. Physiol.* 236, 95-111.

COLE, K.S. (1968). '*Membrane, ions and impulses*.' University of California Press.

COOLEY, J.W. & DODGE, F.A. (1966). 'Digital computer solutions for excitation and propagation of the nerve impulse.' *Biophys. J.* 6, 583-599.

FINKELSTEIN, A. & MAURO, A. (1977). 'Physical principles and formalisms of electrical excitability.' In '*Handbook of Physiology*: ed. 2, Sect. 1, Vol. 1, Pt. 1' (Ed. E.R. Kandel), pp. 161-213. American Physiological Society.

FITZHUGH, R. (1969). 'Mathematical models of excitation and propagation in nerve.' In '*Biological engineering*' (Ed. H.P. Schwan), pp. 1-85. McGraw-Hill.

Ibid. (1973). 'Dimensional analysis of nerve models.' *J. Theoret. Biol.* 40, 517-541.

GOLDMAN, D.E. (1964). 'A molecular structural basis for the excitation properties of axons.' *Biophys. J.* 4, 167-188.

HILLE, B. (1977). 'Ionic basis of resting and action potentials.' In '*Handbook of Physiology*: ed. 2, Sect. 1, Vol. 1, Pt. 1' (Ed. E.R. Kandel), pp. 99-136. American Physiological Society.

Ibid. (1978). 'Ionic channels in excitable membranes: current problems and biophysical approaches.' *Biophys. J.* 22, 283-294.

HODGKIN, A.L. (1939). 'The relation between conduction velocity and the electrical resistance outside a nerve fibre.' *J. Physiol.* 94, 560-570.

Ibid. (1958). 'Ionic movements and electrical activity in giant nerve fibres.' *Proc. Roy. Soc. B* 148, 1-37.

Ibid. (1964). '*The conduction of the nervous impulse*.' Liverpool University Press.

Ibid. (1977). 'Chance and design in electrophysiology: an informal account of certain experiments on nerve carried out between 1934 and 1952.' In '*The pursuit of nature*' A.L. Hodgkin and others), pp. 1-21. Cambridge University Press.

HODGKIN, A.L. & HUXLEY, A.F. (1952*a*). 'Currents carried by sodium and potassium ions through the membrane of the giant axon of *Loligo*.' *J. Physiol.* 116, 449-472.

Ibid. (1952*b*). 'The components of membrane conductance in the giant axon of *Loligo*.' *J. Physiol.* 116, 473-496.

Ibid. (1952*c*). 'The dual effect of membrane potential on sodium conductance in the giant axon of *Loligo*.' *J. Physiol.* 116, 497-506.

Ibid. (1952*d*). 'A quantitative description of membrane current and its application to conduction and excitation in nerve.' *J. Physiol.* 117, 500-544.

HODGKIN, A.L., HUXLEY, A.F. & KATZ, B. (1952). 'Measurement of current-voltage relations in the membrane of the giant axon of *Loligo*.' *J. Physiol.* 116, 424-448.

HODGKIN, A.L. & KATZ, B. (1949). 'The effect of sodium ions on the electrical activity of the giant axon of the squid.' *J. Physiol.* 108, 37-77.

HODGKIN, A.L. & RUSHTON, W.A.H. (1946). 'The electrical constants of a crustacean nerve fibre.' *Proc. Roy. Soc. B* 133, 444-479.

HOYT, R.C. (1963). 'The squid giant axon. Mathematical models.' *Biophys. J.* 3, 399-431.

HUNTER, P.J., McNAUGHTON, P.A. & NOBLE, D. (1975). 'Analytical models of propagation in excitable cells.' *Prog. Biophys. Molec. Biol.* 30, 99-144.

HUXLEY, A.F. (1959). 'Ion movements during nerve activity.' *Ann. N. Yk. Acad. Sci.* 81, 221-246.

JACK, J.J.B. (1976). 'Electrophysiological properties of peripheral nerve.' In '*The peripheral nerve*' (Ed. D.N. Landon), pp. 740-818. Chapman & Hall.

JACK, J.J.B., NOBLE, D. & TSIEN, R.W. (1975)*. '*Electric current flow in excitable cells.*' Clarendon Press.

JAKOBSSON, E. & SCUDIERO, C. (1975). 'A transient excited state model for sodium permeability changes in excitable membranes.' *Biophys. J.* 15, 577-590.

JUNGE, D. (1981). '*Nerve and muscle excitation*' (edn. 2). Sinauer Associates.

KATZ, B. (1966). '*Nerve, muscle and synapse.*' McGraw-Hill.

KEYNES, R.D., GREEF, N.G. & van HELDEN, D.F. (1982). 'The relationship between the inactivating fraction of the asymmetry current and gating of the sodium channel in the squid giant axon.' *Proc. Roy. Soc. B* 215, 391-404.

LIEBERSTEIN, H.M. (1973). '*Mathematical physiology: blood flow and electrically active cells.*' Elsevier.

NEHER, E. & STEVENS, C.F. (1977). 'Conductance fluctuations and ionic pores in membranes.' *Ann. Rev. Biophys. Bioeng.* 6, 345-381.

NOBLE, D. (1973). 'The nerve impulse.' In '*The peripheral nervous system*' (Ed. J.I. Hubbard), pp. 27-45. Plenum.

OFFNER, F.F. (1972). 'The excitable membrane. A physiochemical model.' *Biophys. J.* 12, 1583-1629.

RUSHTON, W.A.H. (1951). 'A theory of the effects of fibre size in medullated nerve.' *J. Physiol.* 115, 101-122.

†ZEEMAN, E.C. (1972). 'Differential equations for the heartbeat and nerve impulse.' In '*Towards a theoretical biology*, 4' (Ed. C.H. Waddington), pp. 8-67. Edinburgh University Press.

†Space beneath the final reference (Zeeman, 1972) permits a postscript on it. In my text I expressed the distrust which most mechanistic biologists feel toward catastrophe theory, but gave the reference which probably least deserves that mistrust. Zeeman would claim to have offered a testable molecular model. Mathematicians, however, may ask themselves whether the testable predictions are really the products of catastrophe theory, or of Zeeman's physical intuition. And biologists will notice at least some respects in which the model conflicts with existing evidence. Together, they may decide whether these latter respects are crucial to the approach.

*A new, paperback printing of Jack, Noble & Tsein, with updated bibliography is due (1983).

The Hodgkin-Huxley axon and the real axon: some recent developments

D. MILLER

The previous paper (Spurway, 1984) summarizes the experimental basis for the Hodgkin-Huxley formalism as a description of the propagated wave of electrical activity (the action potential) in nerve. In this paper I will present a brief review of three aspects of membrane electrical activity where our knowledge has advanced significantly since 1952. As the reader will discover, many developments have served to refine and reinforce rather than refute or replace the physical model of the excitable membrane implied in Hodgkin and Huxley's description. For example, we can now claim to have a better understanding of the mechanisms underlying ion permeation through the membrane, the nature of the membrane potential and of the events preceding and regulating the ionic currents. The power and longevity of the Hodgkin-Huxley scheme reflects the fact that it has provided descriptions which are satisfactory both phenomenologically and mechanistically. The newer insights to be reviewed must be taken into account in any attempt to supercede the Hodgkin-Huxley scheme as a complete model of the electrical activity in the axon. As will be seen, it is the physico-chemical description of the axon which has improved greatly since 1952 and it is with the formalism of this field that the true nature of electrical excitability comes closest to a comprehensive description.

1. SELECTIVITY: THE SPECIFICITY OF THE IONIC CONDUCTANCES IN NERVE

Hodgkin and Huxley's (HH) analysis of the electrical behaviour of the axon hinges upon the idea of distinct, independent conductances for sodium (Na) and potassium (K) ions. The physical counterpart of this concept obviously requires some form of functional or structural connection between the intracellular and extracellular compartments bounding the axonal membrane, permitting the movement of the ions under the electrochemical forces acting upon them. The currents measured in the HH analysis appeared not to saturate as the electrical and chemical driving forces were increased. This implies that the ions are not moved by an intermediating 'carrier' molecule in the membrane, since such systems saturate as the occupancy of the carrier molecule approaches 100%. (It is illuminating to find that Hodgkin and Huxley regarded their empirical description of the conductance changes underlying the action potential to some extent as a failure: their earlier ideas were very much that a 'carrier' model would account for the action potential (see Hodgkin, 1977). Carrier systems are now, however, well understood for a variety of transmembrane ionic movements; the previous article (Spurway, 1984) mentions the Na-K exchange carrier which is ultimately responsible for maintaining the non-equilibrium distribution of these ions which underlies the electrical activity of nerve and muscle fibres.) The maximum rate of transfer of ions approaches 10^7 per sec per active site during the action potential. This is over 100 times greater than the fastest enzyme turnover rate and 10,000 times the rate of the Na-K pump. Thus some form of (approximately) free diffusion pathway through the lipid phase of the axonal membrane is required.

Ion selective channels

Ionic conductances are now generally accepted to be mediated by protein-lined 'pores' penetrating the full thickness of the phospholipid bilayer that constitutes the axonal membrane. These 'pores' or 'channels' have now been characterised in some detail.

The starting point for this work was the observation that a variety of organic and inorganic cations could replace Na and support the action potential, or at least behave like the Na conductance in voltage clamp experiments. In a seminal series of papers, Bertil Hille has reported his investigation of this phenomenon. In 1971 he described the relative permeability of 21 organic cations in frog myelinated nerve fibres. By measuring the reversal potential (i.e. that potential at which no net current flows under voltage clamp) and applying the Goldman equation, (a modification of the Nernst equation described in the previous article (Spurway, 1984 eqn.1 and related text) which takes account of the permeability of all ionic species contributing to the membrane potential), the relative permeability sequence was determined. The organic hydroxylamine ion is almost as permeable (0.94) as sodium, with a series of others of decreasing efficacy (e.g. ammonium 0.16, guanidine 0.13). Significantly, the methylated forms of these substances are virtually impermeable compared with sodium: N-methyl hydroxylamine < 0.056, Methylamine < 0.007, Methyl guanidine < 0.010. Reassuringly, the substances widely used with the aim of replacing sodium osmotically and ionically without supporting conductances through the sodium channel, such as choline (used by Hodgkin and Huxley), Tris(hydroxymethyl)amino methane, and TMA (tetramethylammonium) yield no inward currents. Broadly similar results (though less comprehensive) have been established for the squid axon by other groups.

These results permitted Hille to propose a precise model for the sodium channel (or more correctly, as we shall see later, for that small part of the channel which confers the selectivity upon it; the selectivity filter). He took account of the van der Waals radius of each permeant compound. This gives an idea of the minimum size of orifice that would permit the passage of the ion without the electron orbitals intersecting. At this level the ions can be treated as molecules composed of 'billiard ball' atoms. However, van der Waals size cannot be the only criterion since many of the impermeant ions are also small. Hille argued that the ability to form hydrogen bonds would alter the effective van der Waals size since such bonding permits the closer approach of the bond donor to the acceptor than the simple geometry suggests. A well known example is for ice where the hydrogen of one water molecule overlaps the oxygen of the next by 0.84Å (Pauling, 1960). Thus hydrogen bond donors are effectively smaller when probed by an acceptor such as oxygen than when probed by other atoms. Amino and hydroxyl groups would be 3.0Å in diameter whereas methyl groups would be 3.8Å. Thus an oxygen-lined pore can be designed to combine these steric and chemical specific properties to model the selectivity of the sodium channel. These ideas grew from the comprehensive electrostatic model for cation sensitivity of glass electrodes developed by Eisenmann and his collaborators (see e.g. Eisenmann, 1962).

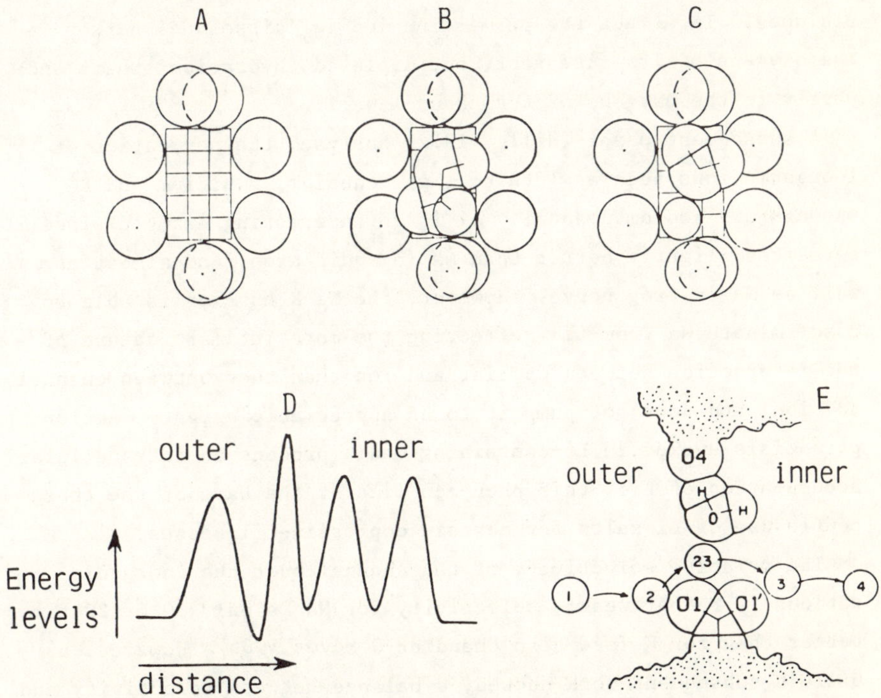

Fig.1 A, B and C represent face-on views of the selectivity filter in the Na-channel. A: is the empty channel; the rings represent oxygen atoms of which 6 are nearly in one plane. The inner grid is 3x5Å. B: includes a hydroxylamine molecule drawn to Van der Waals radii. C: shows a sodium ion and water molecule in transit: In both B and C the overlap represents hydrogen bonding. D: represents four energy barriers encountered by Na in passing the filter. E: illustrates the transit of Na with the filter now in section. Water is hydrogen bonded to oxygen (O4) and sodium passes through states 1-4 corresponding to the energy levels in D (modified after Hille (1971, 1975) by permission of the author and Journal of General Physiology).

Figure 1A (adapted from Hille, 1971 and 1975) shows the shape and size of the selective filter required. Another important feature to emerge (as illustrated in Fig. 1C,E) is that several water molecules can fit in with the permeant ions in passing through the channel: it is energetically highly unlikely that all the water of hydration (that small 'atmosphere' of water molecules which any ion dissolved in water possesses) would be removed during permeation. The oxygen lining to the 'selectivity' filter could be provided by carbonyl, carboxylic acid or phosphate residues. In effect the permeating ion is 'dissolved' partly in the oxygens of its attendant, but depleted, hydration sphere and partly in the oxygens of the channel wall.

A subsequent paper (Hille, 1972) analysed the permeation of inorganic ions such as lithium (Li), caesium, rubidium and K through the sodium channel. Li is an interesting example since it permeates slightly better than Na in squid axon, and almost as well as Na in frog nerve. However, the Na-K carrier is able to discriminate Na from Li (reflecting the more intimate nature of the interaction between carrier and ion than that between channel and ion) and does not pump it to an appreciable extent. Action potentials evoked in Li-containing media produce an intracellular accumulation of Li: this phenomen lies at the base of the therapeutic use of Li salts for certain depressive illnesses.

The relative selectivity of the channel over the inorganic cations is rather weak: selectivity for Na is only 10 or 20 times better than for K (see also Chandler & Meves,1965). However, as Hille's subsequent work showed, a balance between selectivity and energetic cost has been struck.

Hille (1975) has gone on to refine this model and describe the Na channel in terms of Eyring rate theory. The selectivity and permeability characteristics observed can be provided by a channel with four energetic 'barriers': these represent chemical interactions between the permeating ion and the channel's selectivity filter. The energy 'hills' that the penetrating ion must overcome are illustrated in Fig.1D. All the required features can be concentrated in a region only 5Å in length. This allows very much higher flux rates than with a long channel of narrow bore.

The potassium channel Similar methods have been applied with success to the description of the K channel. A lumen 3.0 by 3.5Å is postulated. Rubidium is as permeant as K; caesium and Na are virtually impermeant. The extent of dehydration of K as it traverses the channel is even greater than that of Na in its own channel, (see Hille, 1973, Bezanilla & Armstrong, 1972).

Pharmacological dissection of the action potential

A development crucial to the experimental analysis of the individual ionic conductances was the discovery that the Na and K channels were susceptible to selective blockage by certain drugs and toxins. Tetrodotoxin (TTX) is a highly specific 'poison' of the Na channel in all excitable tissues. It is extracted from the ovary and liver of the Japanese puffer fish or the eggs of one species of salamander. Saxitoxin (STX) is a closely related substance produced by certain shellfish, or more probably the dinoflagellates that live symbiotically with them. These compounds act at minute concentration (nM to μM) and are without detectable effect on the K conductance or resting potential. A similar specificity (though higher dose requirement) is displayed by the quaternary ammonium compounds that block the K conductance. The fact that such specificity is observed reinforces the view that the Na and K conductances are physically represented by channels which are spatially and chemically distinct entities. The use of TTX, for example, permits voltage clamp analysis of the K currents without having to remove sodium ions from the extracellular fluids. The analysis of the voltage dependence and dose-dependence of these blocking actions provides an additional method of probing the precise chemical nature of the Na and K channels. The selectivity filter described earlier probably provides the TTX receptor site: a gaunidinium group in TTX is involved together with one carboxyl group which is susceptible to trimethyloxonium (Hille, 1975).

As a further example of the characterisation of one of the HH processes by such means, the effects of the proteolytic enzyme 'pronase' should be mentioned. This substance removes the inactivation ('h') process from the sodium conductance changes. It only works when applied to the inner face of the axon membrane (see Armstrong, Bezanilla & Rojas, 1973). Again, the insight of the analytical system adopted by Hodgkin and Huxley is vindicated as the Pronase experiments reveal a direct physical counterpart (in this case a protein sub-unit susceptible to this particular enzyme) to the empirical inactivation parameter.

2. SURFACE POTENTIAL AND THE TRANSMEMBRANE FIELD

In the Hodgkin-Huxley description of the axon, the ionic channel conductances are regulated by transmembrane voltage and time. An intriguing observation, well established in the physiological and indeed clinical literature is that nerves and muscles become 'hyperexcitable' even to the point of becoming spontaneously active, when the extracellular or blood plasma calcium concentration ([Ca]) is reduced much below normal. Raising the [Ca^{2+}] has the opposite effect, the so-called 'stabilizing effect' of calcium, rendering cells less easily stimulated and less prone to spontaneous discharge. These effects are shared, usually in lesser degree, by other divalent cations, of which Mg^{2+} is the most relevant physiologically.

If these phenomena are to be explained by anything other than *ad hoc* assumption, then the stabilizing ions must act in a way analogous to voltage and/or time on the ionic conductance channels. This notion is not too unlikely since one can readily conceive that a charged particle, such as the Ca^{2+} ion, might influence the potential field across the membrane. The mechanism underlying such an influence involves the idea, long established in physical chemistry, of the 'double layer' (Gouy-Chapman-Stern, see for example Moore, 1972); where a phase boundary bearing fixed electronic charges exists a potential field develops when mobile charges, such as the ions in an electrolyte solution, are able to redistribute themselves under the influence of the 'fixed' surface charges. The cell membrane and surrounding fluids constitute such a system at both the inner and outer faces. In this case the fixed charges are represented by various chemical species such as amines, phosphates and carboxylic groups carried on the constituent proteins and phospholipids of the biological membrane. A feature of the 'double layer' is that a surface potential exists. Thus the voltage field across the cell membrane is the sum of the 'bulk' potential between the extra- and intracellular fluids, described by the Nernst equation (see the previous article, Spurway, 1984 Equ.1), and the surface potentials, E_o,E_i, at the outer and inner faces of the membrane. E_o or E_i can be derived from the following expression;

$$E_{o,i} = \frac{2(2L\cdot L_{o,i}\cdot C\cdot R\cdot T)^{\frac{1}{2}}}{\sigma}\sinh\left(-\frac{zFE_{o,i}}{2RT}\right) \qquad \text{Equ.1}$$

where L and L_o are the dielectric constants of the membrane and fluid phases, C is the concentration of ions in the bulk electrolyte solution, z is the valency of the ions, and σ is the surface charge density. R,T and F have their usual meaning. This total transmembrane field calculated for typical biological membranes differs considerably from the bulk measurement of the potential difference between intra- and extracellular fluids. Thus, the voltage sensitive channels controlling the electrical activity of the biological membrane are subject to a field (U_m) different to that amenable to simple measurement (V_m). These features are illustrated in Fig.2. A crucial feature, for which there is good evidence, is that the fixed charge density is generally greater on the outer than on the inner surface of the natural cell membrane. Additionally, under normal physiological conditions, the net surface potential is negative on both faces: there is an excess of negatively charged over positively charged chemical groups on the membrane.

Calcium and other ions, can act on this system in two ways: by 'screening' or 'binding'.

Screening The juxtaposition of charged particles (ions) to the surface bearing the fixed charges will tend to 'screen' the electrostatic influence of the latter. This reduces the effective surface charge density and hence surface potential (E_o reduces to E'_o, Fig.2). As a consequence the transmembrane field (U_m) is steepened (U'_m) in the case of the biological membrane, if the ions

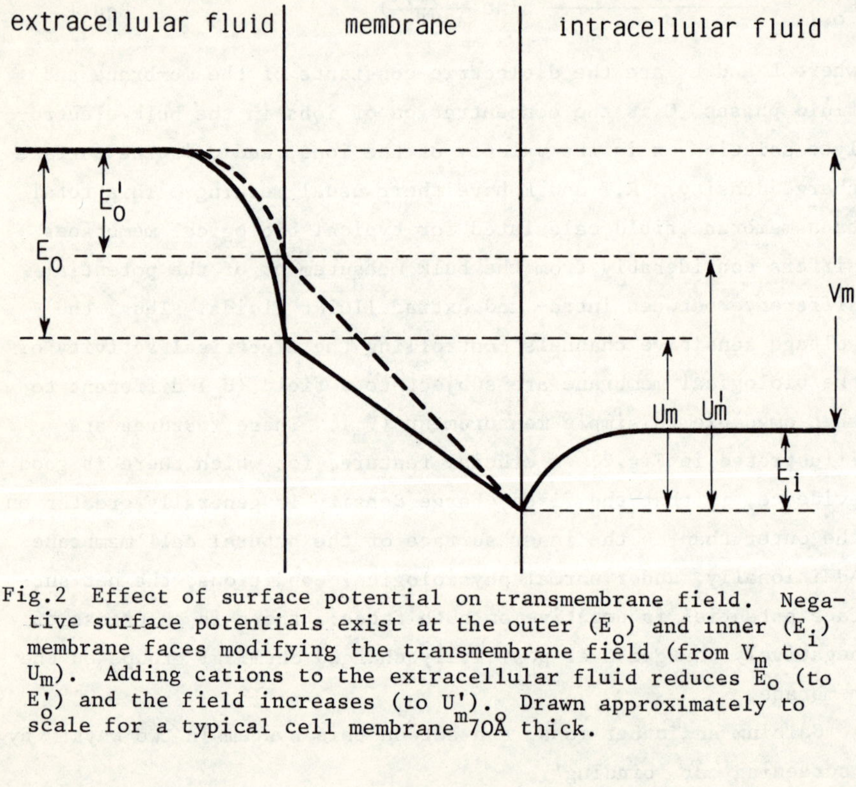

Fig.2 Effect of surface potential on transmembrane field. Negative surface potentials exist at the outer (E_o) and inner (E_i) membrane faces modifying the transmembrane field (from V_m to U_m). Adding cations to the extracellular fluid reduces E_o (to E_o') and the field increases (to U_m'). Drawn approximately to scale for a typical cell membrane 70Å thick.

are applied to the outside surface i.e. if changes are made in the extracellular fluids. However, Vm, the potential difference between bulk intra- and extracellular fluids, does not change (Fig.2). Theoretically, the screening phenomenon is determined only by the valency and sign of the ions concerned (see Equ.1); monovalent ions (K^+, Na^+, Cl^-, HCO_3^- etc.) being least effective, divalent ions (Mg^{2+}, Ca^{2+}, Co^{2+}, Mn^{2+} etc.) more so, and the trivalent ions (La^{3+}, Fe^{3+} etc.) more effective still. The slope of the relationship of screening effect against ion will, however, be Nernstian (see the previous article, Spurway, 1984, Equ.1) e.g. a 29mV shift for a ten-fold change in divalent cation concentration or 19.3mV for trivalent etc. With a charge density of $1/(6.5\text{Å})^2$ (a realistic value) the screening effect of 0.75M monovalent ions (i.e. comparable to that in seawater) is equivalent to the effect of 13.5mM divalent or 250μm trivalent ions. This explains the massive (and often highly toxic) effects that quite small doses of the trivalent ions can have on nerve function.

Binding This describes a (reversible) chemical reaction between the fixed charge (or strictly the chemical group bearing the charge) and the ion concerned. This phenomenon will vary in magnitude with the chemical species concerned. A quantitative description is straightforward in terms of the Law of Mass Action providing that the appropriate 'stability' or 'affinity' constant (the reciprocal of the 'dissociation' constant) for the reaction is known. Thus the binding of Ca^{2+} to a proteinaceous site is typically more avid than that of Mg^{2+}.

The combination of these two processes explains how Ca^{2+} and Mg^{2+} can affect the membrane phenomena to differing degrees depending upon the relative extent of 'screening' and 'binding' for the two ions. The effect can be quantified for example with divalent cations of concentration C^{2+} and affinity K for the binding site:

$$\frac{\sigma}{1+K\left(C^{2+}\exp\left\{\frac{-2FE'_{o,i}}{RT}\right\}\right)} = 2(2L\cdot L_o\cdot C\cdot R\cdot T)\sinh\left(\frac{2FE'_{o,i}}{RT}\right) \qquad \text{Equ.2}$$

(see Pusch, in Luttgau & Glitsch, 1976).

Here σ represents the initial charge density, and $E'_{o,i}$ the modified surface potentials (see Fig.2).

These ideas go a good way to explaining the effects exerted by changes in the extracellular (or intracellular) medium on the behaviour of a variety of electrically excitable tissues. In terms of the Hodgkin-Huxley axon, an increase in $[Ca^{2+}]_o$ will act like a depolarization: the voltage dependent parameters 'm', 'n', and 'h', (see previous article (Spurway, 1984)) are shifted by a few mV for an e-fold change in $[Ca^{2+}]_o$. Quantitative modelling of these phenomena has been carried out in several tissues including squid axon.

The phenomenon of the surface potential has a number of interesting effects: for example, in order to maintain electroneutrality near to the cell membrane, less anions will be present than in the bulk phase. (In effect, the excess of fixed negative charges can be considered to repel the mobile anions free in the bulk solution.) This tends to reduce the 'permeability' calculated for a given concentration of anions present in the bulk solution. This phenomenon can, however, be allowed for in the constant field equation.

These fixed surface charges can also be 'titrated' with hydrogen ions in a manner equivalent to the effects of Mg and Ca described above. Thus an increase in the 'acidity' of the extracellular fluid (a reduction in pH) acts like an increase in $[Ca^{2+}]$. The differential effects of Ca, Mg and H permit a fairly precise chemical characterisation of these surface charges. For example in the node of Ranvier (the electrically active region) of frog nerve three distinct acid residues can be detected with -log dissociation constants (pK_a) of 2.88, 4.58 and 6.28. These groups occur at surface densities of $1/(10.5Å)^2$, $1/(13.2Å)^2$ and $1/(17.3Å)^2$ respectively. The (weak) affinity for Ca ions has a pK of 1.447. With these values and a normal extracellular medium, the surface potential is some -90mV (Hille, Woodhull & Shapiro, 1975).

<u>Diffuse or discrete distribution of surface charge?</u>

One further consideration in the mathematical description of the phenomenon of the surface potential is to what extent the fixed charges can be treated as a 'diffuse' or evenly spread layer at the membrane/electrolyte interface as is described by the Gouy-Chapman-Stern formalism. Since the charges are identifiable as particular chemical groups, weak acids and bases terminating the protein and lipid macromolecules of the membrane, they are clearly spatially discrete entities. The scale of their separation is likely to be the same as that of the 'pores' or 'channels' (see section 1) conducting the ionic flow. Indeed, those charged groups most adjacent to the entrance of such transmembrane conduits are likely to exert a greater electrostatic and possibly steric influence on the ionic flows than those which are more remote. Begenisich (1975) for example has found that the effect of a given $[Ca^{2+}]$ reduction on gK is less than that upon gNa, suggesting a greater negative charge concentration near the Na channels. An exhaustive formal description of the discrete form of charge distribution and surface potential is given by Brown (1974). Fortunately, for most purposes no significant differences between the two types of description emerge from this analysis as being important for the electrically excitable membrane.

3. 'GATING' CURRENTS PRECEDING IONIC CURRENTS

"Much experimental evidence argues against the direct contribution of conventional metabolic sources or ionic gradients in the opening and closing of channels. The only remaining energy source is the electric field. The electric field must do some work on the 'sensor gate' by moving some charges through the membrane field. This work results in a small membrane current reflecting these internal rearrangements. Although this was referred to as a 'carrier current' by Hodgkin and Huxley the term 'gating current' as applied by Hille in the 1970s has been generally adopted." (Hille, 1976 p145)

One concept which developed from the empirical description of the ionic currents of the action potential was of 'gating' particles. These particles were envisaged as dipoles or charged particles residing in the membrane, subject to the transmembrane electric field and having a regulatory effect as a 'switch on' or 'switch off' of the ionic conductances. Hodgkin and Huxley wrote in their description (1952) "the dependence of gNa and gK on membrane potential suggests that the permeability changes arise from the effect of the electric field on the distribution or orientation of molecules with a charge or a dipole moment. The steep dependence of gNa on membrane potential could be understood if such molecules were multivalent a net negative change of 6 would be appropriate, or for a dipole 3 negative changes at one end and 3 positive charges at the other, if they passed through the entire membrane potential field. Such charged particles or dipoles should distribute or orientate themselves in the membrane according to "Boltzmann's Principle". They continued "If such components exist, it is necessary to suppose that their density is relatively low and that a number of sodium ions can cross the membrane at a single active patch. Unless this were true then one would expect the increase in sodium permeability to be accompanied by an outward current comparable in magnitude to the current carried by sodium ions. For movement of any charged particles in the membrane should contribute to the total current and the effect would be particularly marked with a molecule, or aggregate, bearing a large charge."

This notion was put forward by Hodgkin and Huxley as being only one of several possible mechanisms but, predictably, it became adopted into the conceptual framework of the majority of membrane biophysicists. The concept is testable insofar as the movement of such charged particles through the membrane field should produce a current flow. The kinetic analysis provides that the current associated with the 'opening' and 'closing' of the ionic conductance pathway should have a different time-course from that of the ionic current itself. Thus, detection is theoretically possible. Voltage clamp experiments on squid axon were undertaken with the aim of detecting these putative electrical signals of the intramembraneous molecular events. The demonstration of these currents required a precise analysis of the capacity current flowing in the first fraction of a millisecond after the voltage step is imposed upon the membrane. The experiments had to be carefully designed to reduce interference from ionic currents, conventional capacity currents and noise.

i) Ionic currents: The ionic currents carried by Na and K ions and the leakage (time invariant) currents have to be subtracted. The ionic currents can largely be suppressed by replacing the permeant ions with relatively impermeant ones such as Tris (hydroxymethyl) aminomethane to replace Na ions outside, and caesium (Cs) in place of K inside the axon. The highly selective toxin TTX (see section 1) blocks any residual Na current.

ii) Conventional capacity currents: The capacity current can be considered to comprise four components (see Armstrong, 1981);

$$I_{c(total)} = I_{co} + I_{d\ell} + I_{dn} + I_g \qquad \text{Equ.3}$$

where I_{co} is the capacitative flow in a vacuum of the same dimensions as the membrane, $I_{d\ell}$ is the linear (perfect) part of the dielectric charge movement and is larger than I_{co}, I_{dn} is non-linear dielectric current flow not associated with the 'gating' of conductance channels, and I_g is the gating current. The non-linear portion of the capacity current is detected with test pulses of exactly equal size but opposite polarity. This procedure leaves I_{dn}, but it, together with any residual non-linear ionic current can be subtracted on the basis of other criteria.

iii) Noise. Signal averaging over as little as 10 identical sweeps is often sufficient to improve the signal-to-noise ratio and reveal the putative 'gating' currents: under certain circumstances a single sweep can be sufficient.

With these precautions, a small outward (conventional) current has been detected prior to the onset of the sodium current. This current may peak within 100 μs at about 0.1 pA/μm^2. An interesting example of the benefit of hindsight and a vindication, if any be required, if the publication of original experimental records from electrophysiological work, is afforded by a paper by Landowne, Potter and Terrar (1975). These authors re-examined the capacity transients recorded by Hodgkin, Huxley and Katz (1952, their Fig.16) and showed that an assymetry between the capacity transients could be measured, the responses to depolarizing pulses being slightly larger than those to hyperpolarizing pulses. Hodgkin et. al. (1952) had attributed the non-ideal (i.e. non-exponential) decline of the capacity transient to an 'imperfection' in the membrane capacity as described by Curtis and Cole (1938); dielectric loss (see also Fitzhugh & Cole, 1973).

The detailed analysis of the 'gating' currents that has followed since their discovery in 1973 (Armstrong & Bezanilla, 1973, Keynes & Rojas, 1973) at first suggested that the kinetic model of Hodgkin and Huxley would be confirmed. However, more recently a number of details have emerged which have demanded a revision of the classical conceptual framework. These studies continue and, as Armstrong (1981) has candidly reported in his excellent review, the interpretation is still controversial.

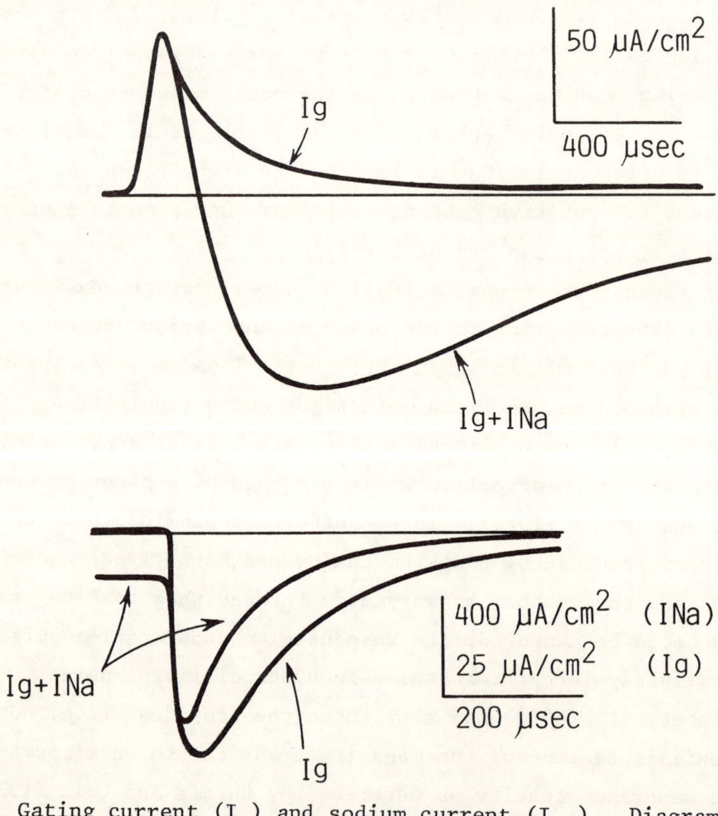

Fig.3 Gating current (I_g) and sodium current (I_{Na}). Diagrams of currents for a depolarizing voltage step (above). With I_{Na} reduced ($[Na^+]_o$ = 5% of normal) I_g is seen as an outward current preceding I_{Na} (Trace $I_g + I_{Na}$). TTX removes I_{Na} completely and reveals I_g in full. The 'tail' current seen on repolarizing (below) declines earlier than I_g, but at a similar rate (see Text for details). Current and time calibrations appropriate for squid axon (modified after Armstrong, 1981).

Fig.3 shows schematically the appearance of the gating current. The upper traces illustrate the currents evoked by a depolarization which would switch on gNa. The capacitative surge of the standard records is eliminated by the subtraction procedure described above and the 'gating' current emerges. As expected this is an outward current (negatively charged particles moving inward) which peaks prior to the onset of the Na current. The complete time course of the gating current is made clearer when the small trace of Na current (these experiments were carried out with 5% of the normal Na concentration) is blocked with TTX. The lower trace shows the 'off' response as the membrane is repolarized, and can be compared to the tail currents of the standard records (Hodgkin-Huxley, 1952). The fact that the 'off'-response gating and Na currents have identical time courses is crucial for the kinetic analysis, as will be discussed later.

The size of the gating current corresponds to about 1500-1900 unit charges per μm^2. Since the density of Na channels is estimated at 330-550 channels per μm^2 (by noise analysis and TTX binding techniques, see section 4) the notional complement of 6 charges per gating unit originally proposed by Hodgkin and Huxley (see the quotation at the head of this section) seems reasonable. The fact that TTX has no effect on the gating current gains further significance since it is thought to act at the outside of the membrane, whereas the activation and inactivation gates are believed to be at the inside of the membrane. (For example, internal application of zinc ions blocks both the gating currents and the Na current. The intracellular effects of pronase on inactivation were described above; see Pharmacological dissection of the action potential in Section 1.)

If the gating currents reflect the behaviour of the gating particles proposed by Hodgkin and Huxley then charge movements will be restricted to a range of voltages between about -70 and +20 mV. As explained in the previous article (Spurway, 1984 Fig.5 and related text), the steepness of the relationship between Na conductance and membrane potential is attributed to the necessity for three (independent) gating particle movements to the 'on' position before the channel becomes patent: gNa is proportional to m^3. The substantial charge movements detected at 'subthreshold' voltages will reflect the movement of particles between various 'closed' states of the channel.

Close examination of the time course has, however, revealed some features which are inconsistent with the m^3h expression for gNa conductance that Hodgkin and Huxley proposed. The gating currents have two distinct components, with a third slower phase emerging at larger depolarizations. A second critical observation, illustrated in Fig.3, is that I_g continues to rise for nearly 100 μsec when the membrane is repolarized whereas I_{Na} falls monotonically: Both these features are at variance with the simple, exponential time-course predicted. The delayed fall of I_g has about the same rate as that of I_{Na} at -60mV, and is only c.1.5 times slower at -90mV. However, the Hodgkin-Huxley m^3 equations predict that I_g would be 3 times slower than the I_{Na} tail.

Armstrong (1981) has proposed a new model for Na channel activation taking in this new information. In essence the scheme proposes 5 closed states (rather than 3) with the first four rate constants fast and the last slow. The charge moved by the faster transitions is half that moved by the slower one. Thus the gating particles are not all identical.

Perhaps the most surprising finding (or non-finding) from the gating current experiments has been the absence of a trace corresponding to the inactivation ('h') process. The Hodgkin-Huxley formalism has 'm' and 'h' as independent parameters. The model proposed by Armstrong couples these two together to the extent that inactivation can only occur when the activation gates have opened; this is readily visualised as a steric blockage of the inner 'mouth' of the channel (where the inactivation gate or particle will fit) by the activation gate when shut. Powerful evidence for this idea is afforded by the observation that the 'off' gating current becomes smaller when inactivation is allowed to develop (i.e. the depolarization is maintained for a few milliseconds). This 'immobilization' of activation gating current is prevented when pronase is used to remove inactivation. The fact that the size of the gating current corresponding to 'm' is

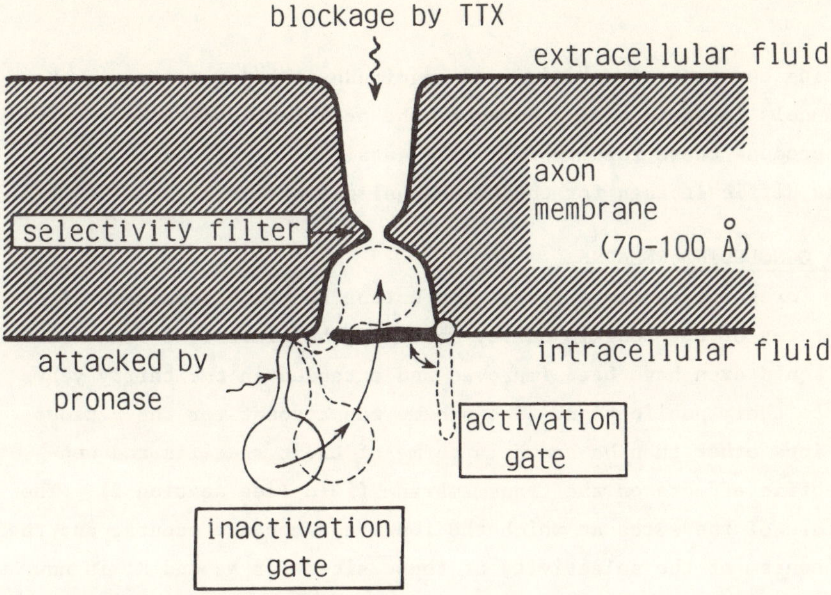

Fig.4 Gating model for the sodium channel in nerve: A schematic, cross-sectional view of the sodium channel. The inactivation can only enter the channel when the activation gate is opened (modified after Armstrong, 1981) See Text for details.

affected by the extent of development of inactivation is, of course, contrary to the Hodgkin-Huxley scheme of 'm' and 'h' independence. Fig.4 illustrates Armstrong's model.

The movement of the charges associated with the conductance gating at the frequencies associated with propagated electrical activity increase the membrane capacity; increased capacity slows propagation as the cable equations show (Spurway, 1984 Equ.4&5). Since such charge movements are an integral part of the sodium conductance which generates the action potential a compromise has to be struck between the need for Na conductance and the minimisation of membrane capacity. Hodgkin (1975) showed, in a typically elegant analysis, that conduction is maximized when capacitance due to sodium gating contributes half the basic membrane capacitance. Since $C_m = 1.0\ \mu F/cm^2$, it is reassuring that gating capacity amounts to $0.4\ \mu F/cm^2$ (Keynes & Rojas, 1974); conduction velocity is thus nearly maximal in the squid giant axon.

Armstrong (see Gilly & Armstrong, 1980) has managed to detect gating currents of a timing and magnitude appropriate for the K channels in squid axon. However, the peculiar conditions required to produce these results will make detailed investigation even more difficult than for the Na-channel gating current.

4. CONCLUDING REMARKS

The foregoing sections reveal the extent to which some central features of the Hodgkin-Huxley model of the electrical behaviour of squid axon have been improved and extended in the thirty years since their publication. The scheme can account for the actions of ions other than Na and K in terms of their specific and non-specific effects on the transmembrane field (see section 2). The nature of the sites at which the ionic conductance occurs, and the mechanism of the selectivity of these sites for Na and K, is now well understood (see section 1). In this area, methods have been developed to study the behaviour of single ionic channels either by statistical analysis of current fluctuations which can be attributed to the behaviour of individual conductance sites (noise analysis: see for example, Neher & Sakmann, 1976, Katz & Miledi, 1972, Stevens, 1972) or by restricting the voltage-clamp to an area of membrane so small that only one or two channels are present in it (see for example: Neher, Sakmann & Steinbach, 1978). The distribution and density of channel occurence can be quantified by radioactive tracer techniques and electrical measurement (e.g. Levinson & Meves, 1975, Keynes & Rojas, 1974). The electrical events reflecting the opening and closing of these conductance sites have also yielded to detection and analysis (see section 3). The principles of the voltage clamp technique and the theoretical framework for understanding and describing electrical excitability as proposed by Hodgkin and Huxley have been extended to many other electrically excitable tissues including skeletal, cardiac and smooth muscle, the synaptic connections from one nerve to another and from nerve to muscle, and secretory tissues as well as nerve fibres from many animal groups including the mammals. Ever increasing complexities have been introduced by the discovery of other ionic permeabilities at work (notably for calcium in many cell types), that current-voltage relations for the individual ionic conductances are frequently non-linear (rectifying), or that the flow of ions may lead to significant accumulation or depletion of the ion concerned. However, in all of this, the central framework of the description in terms of voltage and time dependent conductance changes and the cable equations has been completely vindicated and remains the cornerstone of the biophysical study of electrically active cells.

REFERENCES

Armstrong, C.M. (1981) Sodium channels and gating currents. Physiol. Rev. 61, 644-683.

Armstrong, C.M. & Bezanilla, F. (1973) Currents related to movement of gating particles of the sodium channels. Nature (Lond.) 242, 459-461.

Armstrong, C.M., Bezanilla, F. & Rojas, E. (1973) Destruction of sodium conductance inactivation in squid axons perfused with pronase. J. Gen. Physiol. 62, 375-391.

Brown, R.H. (1974) Membrane surface charge: Discrete and uniform modelling. Prog. Biophys. Mol. Biol. 28, 341-370.

Chandler, W.K. & Meves, H. (1965) Voltage clamp experiments on internally perfused giant axons. J. Physiol. (Lond.) 180, 788-820.

Begenisich, T. (1975) Magnitude and location of surface charges on Myxicola giant axons. J. Gen. Physiol. 66, 47-65.

Bezanilla, F. & Armstrong, C.M. (1972) Negative conductance caused by entry of sodium and cesium ions into the potassium channels of squid axons. J. Gen. Physiol. 60, 588-608.

Curtis, H.J. & Cole, K.S. (1938) Transverse electric impedance of squid giant axon. J. Gen. Physiol. 21, 757-765.

Eisenmann, G. (1962) Cation selective glass electrodes and their mode of operation. Biophys. J. 2, Suppl. 259-323.

Fitzhugh, R. & Cole, K.S. (1973) Voltage and current clamp transients with membrane dielectric loss. Biophys. J. 13, 1125-1140.

Gilly, W.F. & Armstrong, C.M. (1980) Gating current and potassium channels in the squid giant axon. Biophys. J 29, 485-492.

Hille, B. (1971) The permeability of the sodium channel to organic cations in myelinated nerve. J. Gen. Physiol. 58, 599-619.

Hille, B. (1972) The permeability of the sodium channel to metal cations in myelinated nerve. J. Gen. Physiol. 59, 637-658.

Hille, B. (1973) Potassium channels in myelinated nerve: selective permeability to small cations. J. Gen. Physiol. 61, 669-686.

Hille, B. (1975) Ion selectivity, saturation and block in sodium channels. A four-barrier model. J. Gen. Physiol. 66, 535-560.

Hille, B. (1976) Gating in sodium channels of nerve. Annu. Rev. Physiol. 38, 139-152.

Hille, B., Woodhull, A.M., & Shapiro, B.I. (1975) Negative surface charge near sodium channels of nerve: divalent cations, monovalent ions, and pH. Phil. Trans. R. Soc. B. 270, 301-318.

Hodgkin, A.L. (1975) The optimum density of sodium channels in an unmyelinated nerve. Phil. Trans. R. Soc. (B) 270, 297-300.

Hodgkin, A.L. (1977) Chance and design in electrophysiology: an informal account of certain experiments on nerve carried out between 1934 and 1952. In 'The Pursuit of Nature' (A.L. Hodgkin and others) pp 1-21 Cambridge University Press.

Hodgkin, A.L. & Huxley, A.F. (1952) A quantitative description of membrane current and its application to conduction and excitation in nerve. J. Physiol. (Lond.) 117, 500-544.

Hodgkin, A.L., Huxley, A.F. & Katz, B. (1952) Measurement of current-voltage relations in the membrane of the giant axon of Loligo. J. Physiol. (Lond.) 116, 424-448.

Katz, B. & Miledi, R. (1972) The statistical nature of the acetylcholine potential and its molecular components. J. Physiol. (Lond.) 224, 665-699.

Keynes, R.D. & Rojas, E. (1973) Characteristics of the sodium gating current in the squid giant axon. J. Physiol. (Lond.) 233, 28P-30P.

Keynes, R.D. & Rojas, E. (1974) Kinetics and Steady-state properties of the charged system controlling sodium conductance in squid giant axon. J. Physiol. (Lond.) 239, 393-434.

Landowne, D., Potter, L.T., & Terrar, D.A. (1975) Structure function relationships in excitable membranes. Ann. Rev. Physiol.

37, 485-508.

Levinson, S.R. & Meves, H. (1975) The binding of tritiated tetrodotoxin to squid giant axons. Phil. Trans. R. Soc. B. 270, 349-352.

Lüttgau, H.Ch. & Glitsch, H.G. (1976) Membrane physiology of nerve and muscle fibres. Fortschr. Zool. 24, 1-132.

Moore, W.J. (1972) 'Physical Chemistry' pp 510-513, 5th Edn., (Longmans, London).

Neher, E. & Sakmann, B. (1976) Noise analysis of drug-induced voltage clamp currents in denervated frog muscle fibres. J. Physiol. (Lond.) 258, 705-729.

Neher, E., Sakmann, B. & Steinbach (1978) The extracellular patch clamp: a method for resolving currents through individual open channels in biological membranes. Pflügers Arch. 375, 219-228.

Pauling, L. (1960) 'Nature of the Chemical Bond'. Cornell University Press, Ithaca, N.Y. (3rd Edn.) 449-504,514.

Spurway, N.C. (1984) The physiology of the nerve action potential - A guide for the perplexed. In 'Mathematics in Medicine & Biomechanics' ed. Eason, G., Shiva.

Stevens, C.F. (1972) Inferences about membrane properties from electrical noise meansurements. Biophys. J. 12, 1028-1047.

The Fluid mechanics of blood

E.A. TROWBRIDGE

1. INTRODUCTION

The popular conceptual view of blood flowing in arteries and arterioles which is presented in the literature is of individual separated particles, surrounded by a continuous plasma fluid, being convected along in a laminar flow. As a consequence the volumetric flow rate of erythrocytes is taken as HQ, where H is the blood haematocrit and Q is the volumetric flow rate of the fluid.

The particulate suspension theory of blood is based on the following assumptions:

Assumption 1 The erythrocyte is surrounded by a continuous fluid at all times.

Assumption 2 The erythrocyte moves with the ambient fluid locally.

Assumption 3 The presence of other erythrocytes does not disturb the continuum flow in a local neighbourhood of a single particle.

Assumption 4 The linear dimensions of the erythrocytes are so small that a point concentration (haematocrit) can be defined in the blood.

Assumption 5 The particulate suspension theory specifically excludes any transference of momentum from erythrocyte/erythrocyte collisions. That is, the effects arising from erythrocyte/erythrocyte collisions are always ignored.

Confirmation of these assumptions can be found in many places in the literature. For example, Happel and Brenner (1965) discuss both dilute and concentrated suspensions. Batchelor (1967) gives a detailed hydrodynamic account of the viscosity of particulate suspensions. Batchelor and Green (1972) investigate the bulk stress of concentrated suspensions.

If any of the assumptions which are stated above are found to be invalid for blood then particulate suspension theory cannot provide an adequate realistic description of this fluid. [The same remarks apply to micropolar theories of blood, such as those proposed by Kang and Eringen (1976). These theories make similar assumptions to those above but provide some 'structure' at a fluid point by the introduction of a triad of deformable directors].

In reality for blood at normal haematocrits (40% say) experimental evidence suggests that not one of the five assumptions, listed above, is valid.

Reference to reports of the high-speed cinephotography of blood flow, see for example Bloch (1966), Goldsmith and Karino (1980) and Thomas et al. (1965) reveals that:

1. An individual red cell is in frequent collision (often multi-body collisions) with other red cells. [So that Assumption 5 not valid.]

2. If there are frequent collisions between red cells the erythrocytes are not surrounded by a continuous fluid at all times. [So that Assumptions 1 and 3 are not valid].

3. The erythrocytes tumble and roll and follow paths which are not parallel to the tube walls. The erythrocytes have substantial radial velocities and are not convected along the plasma streamlines. [So that Assumption 2 not valid]. The reason for the invalidity of Assumption (4) can be found in Trowbridge (1982a).

In such circumstances particulate suspension (which includes micropolar fluids) theories may provide highly sophisticated mathematics but they will give no insight into the true rheological behaviour of blood. Even the theories which purport to include particle interactions (see Batchelor and Green (1972), for example) always assume that erythrocytes never collide. Indeed such an occurrence in the suspension theories would lead to infinite singularities. As a consequence an interpretation of clinical screening tests such as erythrocyte sedimentation rates (E.S.R) are based on theories (see Caro et al. (1978)) whose assumptions are totally unrealistic.

The axial streaming concept of erythrocytes (Assumption 2) proliferates in the literature, (see, for example, Thomas (1962), Fung (1977), Rosenblum (1977), Meiselman (1980), Cokelet (1976), Fung (1981)) even though high-speed photographic evidence suggests that this is totally unrealistic as well. The experimentally measured velocity profiles of blood are presented in a way which imply that the red cell move parallel to the cylinder walls with the ambient fluid, for example, Goldsmith and Karino (1980), even though the same scientists present evidence that the individual erythrocytes can have substantial radial velocities.

The ability to observe the individual red cells by high-speed photography (say) and spatial magnification has given the erythrocyte a special status in blood flow. The preoccupation with the red cells as an individual particle has prohibited the development of a physical theory of arteriole blood flow. An attempt to include the complicated microstructural behaviour of blood in a realistic way leads to intractable mathematics. As a consequence mathematical problems have been solved which lie outside the range of validity for which the model was constructed.

Visual evidence suggests, however, that a natural approximation of blood is a fluid continuum. This statement relies on an understanding of the fundamental physical principles which lead to the continuum hypothesis. In the classical investigations of air and water the molecules are sufficiently small and the macroscopic lengths sufficiently large to enable the choice of a continuum length scale giving many molecules in a small spatial volume at any instant. Averages are taken over space alone to obtain the continuum point density and velocity and the question of "What is an instant" is never raised. With fluids composed of particles of vastly differing weight and size the answer to the above question is of paramount importance.

The high speed photography of Bloch (1966), for example, points to the answer. When arteriole blood flow is viewed using photography at a framing rate of 8000 frames/sec. the red cell motion is essentially 'stopped'. The time interval of 1/8000 sec. is insufficient to register a change in position of a red cell on the photographic frame at the spatial magnification used. The red cells are 'seen' as particles. Yet at the same spatial magnification the whole blood flow is seen as a continuum using the human sensor, the eye, alone. The particulate nature of the red cells as well as the plasma has been 'smoothed out'.

The reason for this is the framing rate adopted by the human sensor. This is approximately 1/24 sec. The individual positions taken up by the red cells throughout the time interval of 1/24 sec. cannot be identified as individual events. They are blurred over a region of space, much as an arm is blurred when it is moved during a photograph taken by a camera with normal shutter speed. Consequently the human sensor is left with the impression of a red fluid continuously distributed over space moving with a 'bulk' velocity. The red cell contributes to this continuum just as much as the plasma.

This article develops a physical theory based on these ideas. The different characteristics and behaviour of the red cells in health and disease are mirrored in different values of the phenomenological constants which arise from the theory. These parameters are, of course, an averaged and therefore approximate description of the blood characteristics. However, since most interpretations of blood flow are based on experiments which use this averaging (phenomenological) procedure (for example, in the evaluation of fluid viscosity) the theory should be of value in explaining observed phenomena.

In a continuum approximation the whole particulate nature of blood is continuously distributed throughout space and time but the importance of the relative length and time scales, as described above, must always be remembered. [For further details see Trowbridge (1982a)].

The continuum point density, ρ, of whole blood at a continuum instant is obtained by the method explained above. In general both red cell mass and plasma will be present at a point during an instant. As a consequence the plasma will have a continuum point density $\rho^{(1)}$ and the erythrocytes a point density $\rho^{(2)}$ such that

$$\rho = \rho^{(1)} + \rho^{(2)}. \qquad (1)$$

In equilibrium the two constituents of whole blood may be considered as incompressible in the following sense. When the plasma and erythrocytes are separated, by centrifuge techniques say, the two constituents have, in isolation, constant densities $\rho_I^{(1)}$ and $\rho_I^{(2)}$ respectively. If a mass m_1 of plasma occupies a volume V_1 then $\rho_I^{(1)} = \frac{m_1}{V_1}$ and if a mass m_2 of erythrocytes occupies a volume V_2 then $\rho_I^{(2)} = \frac{m_2}{V_2}$. The incompressibility condition assumes that there is no change in volume when the whole blood mixture is formed in equilibrium, so that a mass of blood $m = m_1 + m_2$ occupies a volume $V = V_1 + V_2$. The point density of plasma in the blood mixture is $\rho^{(1)} = \frac{m_1}{V_1 + V_2}$ and that of erythrocyte is $\rho^{(2)} = \frac{m_2}{V_1 + V_2}$.

It follows that in equilibrium

$$\frac{V_1}{m_1}\frac{m_1}{V} + \frac{V_2}{m_2}\frac{m_2}{V} = 1$$

so that

$$\frac{\rho^{(1)}}{\rho_I^{(1)}} + \frac{\rho^{(2)}}{\rho_I^{(2)}} = 1 . \tag{2}$$

Using equations (1) and (2) it follows that in equilibrium

$$\rho^{(1)} = \frac{\rho_I^{(1)}(\rho - \rho_I^{(2)})}{\rho_I^{(1)} - \rho_I^{(2)}}, \quad \rho^{(2)} = \frac{\rho_I^{(2)}(\rho - \rho_I^{(1)})}{\rho_I^{(2)} - \rho_I^{(1)}} . \tag{3}$$

In dynamic flows when <u>one constituent is moving relative to the other</u>, from a continuum viewpoint, it is not possible to assume that the incompressibility condition (2) holds. Equation (1) is still valid however.

The mass average or linear momentum of the mixture is given by

$$\rho \underset{\sim}{v} = \rho^{(1)} \underset{\sim}{v}^{(1)} + \rho^{(2)} \underset{\sim}{v}^{(2)} \tag{4}$$

where $\underset{\sim}{v}^{(1)}$ and $\underset{\sim}{v}^{(2)}$ are the continuum velocities of the plasma and erythrocytes. Equation (4) states that the whole blood linear momentum is the sum of the linear momenta of the individual constituents.

2. BLOOD IN EQUILIBRIUM

To develop a physical theory of blood flow it is necessary to consider initially the mechanical equilibrium of the fluid.

For equilibrium of a fluid with constant density

$$\rho \underset{\sim}{F} = \underset{\sim}{\nabla} p \tag{5}$$

where $\underset{\sim}{F}$ is the total body force acting on the fluid and p is the static-fluid pressure. The pressure is usually considered as the negative value of the average of the normal stresses and is isotropic in the sense that all orthogonal axes of reference are principal axes for the stress tensor. If the body force is conservative then a potential function ϕ can be introduced and

$$-\rho \underset{\sim}{\nabla} \phi = \underset{\sim}{\nabla} p. \tag{6}$$

For example in the Earth's gravitational field $\phi = -gz$, where z is measured from the fluid free surface. For a fluid of uniform density

$$p = p_A + \rho g z, \qquad p_A = \text{atmospheric pressure}. \tag{7}$$

For a fluid, such as blood, which is assumed to consist of two different species the density will not be uniform in equilibrium. For an initial model the equation of state of whole blood will be taken as

$$p = p_o + k\rho ,$$

where k may depend on temperature. This is simply a linear approximation of the more general statement $p = f(T, \rho)$.

In a continuum mixture theory, for whole blood

$$\rho \underset{\sim}{F} = \underset{\sim}{\nabla} p \tag{8}$$

for plasma

$$\rho^{(1)} \underset{\sim}{F}^{(1)} = \underset{\sim}{\nabla} p^{(1)} \tag{9}$$

and for erythrocytes

$$\rho^{(2)} \underset{\sim}{F}^{(2)} = \underset{\sim}{\nabla} p^{(2)} , \tag{10}$$

where the total body force

$$\rho \underset{\sim}{F} = \rho^{(1)} \underset{\sim}{F}^{(1)} + \rho^{(2)} \underset{\sim}{F}^{(2)} , \tag{11}$$

is the sum of the total body forces acting on the plasma and erythrocyte constituents.

The blood pressure is the sum of the erythrocyte and plasma partial pressures

$$p = p^{(1)} + p^{(2)} . \tag{12}$$

If the body forces are considered as conservative and the pressures are assumed to be proportional to the constituent densities then equations (9), (10) and (11) become

$$- \rho \underset{\sim}{\nabla} \phi = k \underset{\sim}{\nabla} \rho \tag{13}$$

$$- \rho^{(1)} \underset{\sim}{\nabla} \phi^{(1)} = k \underset{\sim}{\nabla} \rho^{(1)} \tag{14}$$

and

$$- \rho^{(2)} \underset{\sim}{\nabla} \phi^{(2)} = k \underset{\sim}{\nabla} \rho^{(2)} . \tag{15}$$

For a unidirectional field it follows that

$$\rho = \rho(h) \exp \left[- \left\{ \frac{\phi(x) - \phi(h)}{k} \right\} \right] \tag{16}$$

$$\rho^{(1)} = \rho^{(1)}(h)\exp\left[-\left\{\frac{\phi^{(1)}(x) - \phi^{(1)}(h)}{k}\right\}\right] \qquad (17)$$

$$\rho^{(2)} = \rho^{(2)}(h)\exp\left[-\left\{\frac{\phi^{(2)}(x) - \phi^{(2)}(h)}{k}\right\}\right] \qquad (18)$$

where x = h is the point at which the blood mixture changes to a single constituent erythrocyte fluid. In particular in the Earth's gravitational field where x is measured from the Earth's surface $\phi^{(2)} = + gx$ and equation (18) becomes

$$\rho^{(2)} = \rho_I^{(2)} \exp\left(-\frac{g(x - h)}{k}\right) . \qquad (19)$$

Furthermore equation (3) gives

$$\rho = \rho_I^{(1)} + (\rho_I^{(2)} - \rho_I^{(1)}) \exp\left(-\frac{g(x - h)}{k}\right) \qquad (20)$$

and

$$\rho^{(1)} = \rho_I^{(1)} \left[1 - \exp\left(-\frac{g(x - h)}{k}\right)\right] . \qquad (21)$$

There are two types of interactions between 'molecules' for blood in equilibrium. The interactions, both kinetic and potential in a purely mechanical theory, between like molecules, that is erythrocyte/erythrocyte and plasma/plasma interactions are balanced by the introduction of the normal stresses, the partial pressures p_1 and p_2. There are also interactions between the erythrocytes and the plasma molecules. These are balanced by the correct interpretation of the body force potentials $\phi^{(1)}$, $\phi^{(2)}$ and ϕ. The Earth's gravitational pull provides an external body force acting on the whole system. The potential field created by the presence of the mass of the erythrocyte constituent which acts on the plasma cannot be neglected as in dilute solutions. Since the ratio of the mass of a single red cell to that of an average plasma molecule (based on a molecular weight of 78,000) is approximately 10^{11}, the 'gravitational force' exerted on the plasma by the erythrocyte fluid is much greater than that exerted on the erythrocyte fluid by the plasma. [It is not correct to think of the red cells as particles in a viscous fluid (see Introduction)]. As a consequence the dominant potential field acting on the erythrocytes is the Earth's gravity which explains why $\phi^{(2)}$ was expressed in the form $\phi^{(2)} = gx$. The plasma however is influenced not only by the gravitational pull but also by the presence of the erythrocytes, $\phi^{(1)}$ <u>cannot</u> be expressed simply as $\phi^{(1)} = gx$ as in a single constituent fluid. In fact equations (17) and (21) imply that $\phi^{(1)} \to \infty$ as $x \to h$ so that plasma is excluded from the base of a vertical capillary that contains blood in equilibrium. This feature is observed experimentally (see Trowbridge (1982a)). The whole blood potential field cannot be taken as $\phi = gx$ either. Equations (16) and (21) imply that

$$\phi(x) - \phi(h) = -k \log_e \left[\frac{\rho_I^{(1)} + (\rho_I^{(2)} - \rho_I^{(1)}) \exp\left[-g\frac{(x-h)}{k}\right]}{\rho_I^{(2)}} \right].$$

As $x \to h$, $\phi(x) \to \phi(h)$ and the above equation is identically satisfied.

Equation (19) states that the density of the erythrocyte fluid approaches zero for large values of x. Effectively $\rho^{(2)} \to 0$ when $x - h = \frac{10k}{g}$. In equilibrium the blood mixture changes from red cells only at the base of a vertical tube to pure plasma at a point determined by the parameter k. The value of k can be determined when the haematocrit, H, of the blood sample and the length L, of tubing containing the sample is known. The total mass in the system is invariant so that

$$\int_0^L \rho \, dx = \bar{\rho} L = \text{mass of blood sample}$$

where $\bar{\rho}$ is the blood density when the erythrocytes are uniformly distributed.

Consequently equation (20) implies that

$$\exp\left(\frac{gh}{k}\right)\left[1-\exp\left(-\frac{gL}{k}\right)\right]=\left[\frac{\bar{\rho}-\rho_I^{(1)}}{\rho_I^{(2)}-\rho_I^{(1)}}\right]\frac{gL}{k} \quad . \qquad (22)$$

Equation (19) is a continuum mixture analogue of the Boltzmann distribution found in statistical mechanics. The constant k is a measure of the contribution to the absolute pressure of the thermal energy in the system. The source of energy which produces 'molecular motion' of the erythrocytes is the kinetic energy of the plasma molecules.

Equations (20) and (21) also show that whole blood and plasma tend to the plasma isolation density for large x.

3. THE SEDIMENTATION OF BLOOD

When a sample of blood (suitably treated to prevent coagulation) is allowed to stand in a vertical tube the erythrocytes are observed to sediment under the action of the Earth's gravitational field.

The classical approach to this problem would treat the red cells as suspended particles. The sedimentation rate is usually calculated by considering the viscous drag exerted on the cell as it moves through the plasma under the influence of the Earth's gravitational pull. This theory is only a good model of reality when the haematocrit is infinitesimally small (one red cell). It provides a reasonable approximation when the haematocrit is very low and the Assumptions (1-5) stated in the introduction are valid. For normal blood haematocrits (40% say) frequent red cell collisions occur, the Assumptions (1-5) are not valid and a particulate suspension theory of sedimentation cannot possibly describe the erythrocyte sedimentation rate.

The mixture theory approach adopted in this paper would not single out the red cell for special treatment. The observed sedimentation of red cells is simply the fluid system trying to attain the equilibrium distribution described in the previous section. As the red cells move from their initial uniform distribution they will be subject to collisions with plasma molecules and other erythrocytes. The progress will be erratic, the red cells actual velocity being determined by the collisions they experience. (The theory presented here is purely mechanical, any electrical or chemical effects are not included although they may play an important part in the sedimentation of blood in normal plasma. For example, the possibility of rouleaux formation is very important. When rouleaux form it would appear that the isolation density $\rho_I^{(2)}$ increases. This has a profound effect both on sedimentation rates and the behaviour of blood flowing through arterioles).

From a continuum viewpoint the linear momentum of the whole blood will be zero, there is no flow present. The plasma constituent simply moves relative to the red cell constituent while the mass centre of the infinitesimal volume which represents the material blood mixture continuum particle does not change its position.

When plasma molecules collide with plasma molecules there is a transference of momentum. In a continuum theory, when no flow is present, this momentum transfer would be balanced by the introduction of a normal stress. In a similar manner collisions of red cells with red cells can be balanced by a normal stress component related to the erythrocyte constituent. These are the partial pressures used in the previous section when no bulk flow is present.

As the erythrocytes and plasma molecules are adjusting to their own particular equilibrium distributions, collisions between red cells and plasma molecules take place. A measure of the ability of the molecular motion to eliminate any non-uniformity in mass distribution will be referred to as the continuum 'diffusion coefficient' D. This constant is a phenomenological constant which can be determined by experimentation in a non-gravitational field. It is not half the mean square distance travelled by a particle in a time interval, divided by the time interval. The latter evaluation of D is only strictly true for a single particle, is approximately true for very dilute blood mixtures and is not valid for blood mixtures when erythrocyte-erythrocyte interactions are important.

Since the erythrocytes and plasma constituents are moving with different continuum velocities the transfer of momentum between plasma molecules and red cells must be balanced by the introduction of a body force in the continuum theory which acts as an internal force for the system as a whole. It is considered as a body force since the presence of the plasma is felt equally by the total erythrocyte constituent. There will be an equal and opposite force acting on the plasma due to the presence of the erythrocytes. The form of this body force will be that used in previous discussions of mixtures with two constituents (see Adkins (1963) and Trowbridge (1982a)).

It is assumed that the 'diffusive' or 'sedimentation' drags can be written as

$$\rho^{(2)}\underset{\sim}{\psi}^{(2)} = \frac{k}{D}\,\frac{\rho^{(1)}\rho^{(2)}}{\rho}\,[\underset{\sim}{v}^{(1)} - \underset{\sim}{v}^{(2)}] = -\rho^{(1)}\underset{\sim}{\psi}^{(1)} \qquad (23)$$

where $\rho^{(2)}\underset{\sim}{\psi}^{(2)}$ is the body force acting on the erythrocytes due to the presence of the plasma. The body force $\rho^{(1)}\underset{\sim}{\psi}^{(1)}$ is that exerted on the plasma due to the presence of the erythrocytes.

The equations governing the sedimentation of erythrocytes in the Earth's gravitational field are,

<u>Conservation of mass</u>

$$\frac{\partial \rho^{(2)}}{\partial t} + \frac{\partial}{\partial x}\,(\rho^{(2)}v^{(2)}) = 0 \qquad (24)$$

where $\rho^{(2)}v^{(2)}$ is the continuum linear momentum of the erythrocyte constituent.

<u>Conservation of linear momentum</u>

$$-\rho^{(2)}\frac{d\phi^{(2)}}{dx} - \frac{\partial p^{(2)}}{\partial x} + \frac{k}{D}\,\frac{\rho^{(1)}\rho^{(2)}}{\rho}\,(v^{(1)} - v^{(2)}) = 0 \qquad (25)$$

where $\rho^{(1)}v^{(1)}$ is the linear momentum of the plasma constituent.

With the absolute pressure $p^{(2)} = p_o^{(2)} + k\rho^{(2)}$ and $\phi^{(2)} = -gx$ equation (25) reduces to

$$\rho^{(2)} g - k \frac{\partial \rho^{(2)}}{\partial x} - \frac{k}{D} \rho^{(2)} v^{(2)} = 0 \tag{26}$$

where equations (1) and (4) have been used with $\rho v = 0$.

The equations above must be interpreted as the movement of a 'red fluid' through a 'plasma atmosphere' under the influence of the Earth's gravitational pull. The plasma exerts a resistive body force on the 'red fluid' (the erythrocytes) while gravity 'pulls' the fluid vertically downwards. Conceptually there is little difficulty in imagining the erythrocytes as a 'sticky' red fluid in isolation flowing through a vertical tube when air molecules are interspersed between the red cells. Movement of red cells through plasma is just the same except that the interspersion is now with plasma molecules. The air molecules provide little resistance compared with the plasma, in the sense that the transference of momentum when an air molecule strikes a red cell is much less than that which occurs when a plasma molecule collides with a red cell.

The movement of the red cells is so slow during sedimentation that the inertial component of the red fluid in equation (25) has been neglected.

A combination of equations (24) and (26) shows that $\rho^{(2)}$ satisfies the equation

$$\frac{k}{D} \frac{\partial \rho^{(2)}}{\partial t} = k \frac{\partial^2 \rho^{(2)}}{\partial x^2} - g \frac{\partial \rho^{(2)}}{\partial x} . \tag{27}$$

If it is thought that sedimentation is the dominant mass transfer process then by taking the characteristic length associated with the system as the length, L, and by taking the characteristic time as $\frac{Lk}{gD}$, equation (27) may be non-dimensionalised to give

$$\frac{\partial\hat{\rho}^{(2)}}{\partial\hat{t}} = \frac{k}{gL}\frac{\partial^2\hat{\rho}^{(2)}}{\partial\hat{x}^2} - \frac{\partial\hat{\rho}^{(2)}}{\partial\hat{x}} \tag{28}$$

where ^ denotes a dimensionless variable. The dimensionless parameter $\frac{k}{gL}$ can be thought of as the ratio of two velocities $\frac{D}{L}\Big/\frac{Dg}{k}$. Interpreting the parameter Dg/k as the sedimentation velocity and D/L as the diffusion velocity, it follows that $\frac{k}{gL} << 1$ when sedimentation is dominant. The characteristic time, $\frac{Lk}{gD}$, is then the time it takes for the erythrocyte constituent to sediment through a length L.

When $\frac{k}{gL} << 1$ equation (34) becomes, in dimensional form

$$\frac{\partial\rho^{(2)}}{\partial t} + \frac{Dg}{k}\frac{\partial\rho^{(2)}}{\partial x} = 0. \tag{29}$$

In this limit the erythrocyte constituent moves through the plasma all at the same speed. The erythrocyte boundary is sharply defined and travels as a plane with speed $\frac{Dg}{k}$. The efforts of the molecular motion to eliminate the non-uniformity of matter, caused by sedimentation, modulate this behaviour.

In passing, it should be mentioned that if diffusion is thought to be the dominant process of mass transfer than L^2/D should be taken as the characteristic time and then equation (27) non-dimensionalised becomes

$$\frac{\partial\hat{\rho}^{(2)}}{\partial\hat{t}} = \frac{\partial^2\hat{\rho}^{(2)}}{\partial\hat{x}^2} - \frac{gL}{k}\frac{\partial\hat{\rho}^{(2)}}{\partial\hat{x}}. \tag{30}$$

In this case D/L >> Dg/k so that gL/k << 1 and equation (30) becomes in dimensional form

$$\frac{\partial\rho^{(2)}}{\partial t} = D\frac{\partial^2\rho^{(2)}}{\partial x^2}$$

the classical Fickian diffusion equation.

The parameter D/k plays an important part in the behaviour of blood flowing through arterioles (see Trowbridge (1982b)). It is a universal constant for a specific blood mixture at a given temperature. It can be identified as the erythrocyte sedimentation rate in an infinitely long tube produced by a unit gravitational field (i.e. g = 1). The parameter D/k will be called the 'sedimentation constant' since it is similar to the sedimentation constant observed by Svedberg in centrifugal fields (see Moore (1962), Pg 765). The dimensions of D/k are time. E.S.R. experiments and values obtained from blood flow experiments using this theory suggest that, for blood mixtures, the basic unit should be 10^{-8} sec. It can be measured in the laboratory, to a good approximation, by using the Earth's gravitational field and could be used in future for cataloguing different blood mixtures in disease. For use in the theory of normal arteriole blood flow, see Trowbridge (1982b), sedimentation rates should be measured in preparations for which red cells retain their individuality.

4. ARTERIOLE BLOOD FLOW

By considering blood as a mixture of two fluids, a 'red' erythrocyte fluid and a 'yellow' plasma fluid a mechanical theory of blood flow can be constructed for vessels whose diameters are in the range 20μ-1000μ.

Erythrocyte/erythrocyte interactions are balanced by the introduction of erythrocyte stress components. The plasma molecule/plasma molecule interactions are balanced by plasma stress components. The sum of the erythrocyte stress tensor and the plasma stress tensor provides the blood mixture stress tensor. It is found (see Trowbridge (1982b)) that an assumption of a linear fluid constitutive equation for each of the ingredients in the blood mixture is sufficient to explain the many anomalies found in arteriole blood flow. Erythrocyte/plasma interactions are balanced by the diffusive body force introduced in the section on blood sedimentation. It can also be shown (see Trowbridge (1982b)) that the interaction between a solid boundary and the two constituents in the mixture is different in arteriole blood flow. Although an arteriole wall (or equivalently a glass cylinder) exerts a normal force on both constituents the region of influence of the arteriole wall is different for each constituent. It is this feature which provides the cell free layer. In reality this layer is simply an illusion which is created by the crudity of the observer's time scale. The red cells actually contact the wall but 'smoothing' over space and time (see Introduction) gives the impression of the absence of red cells close to the wall. It is the presence of this layer in the continuum mixture theory which enables the observed flow properties of blood to be predicted.

A new dimensionless similarity parameter, which provides a measure of the relative importance of the two frictional forces (shear stress and 'diffusive' body force), arises naturally from the mixture theory. The size of this parameter indicates that the drag of the erythrocytes on the plasma and the concomitant 'pull' of the plasma on the erythrocytes is much more important than the viscous drags experienced by either fluid. It is this feature which dominates the flow and provides the mechanism for the high viscosities of blood compared with water.

Details of the continuum mixture theory of arteriole blood flow and its use in explaining the Fahraeus effect, the Fahraeus-Lindqvist effect, the relationship between feed, tube and discharge haematocrit,observed experimentally by Barbee and Cokelet (1971) and the radial viscosity profiles observed by Bugliarello et al (1965) can be found in Trowbridge (1982b) and Trowbridge and Meadowcroft (1982).

Acknowledgements:

The author would like to acknowledge the assistance of Jill Bradley in the preparation of this chapter. He would also like to thank the organisers of the Strathclyde University Workshop for their invitation to a stimulating week of scientific discussion. The interdisciplinary contacts created in this week should prove invaluable in the future.

REFERENCES

ADKINS, J.E. (1963) Non-linear diffusion. 1. Diffusion and flow of mixtures of fluids. Phil. Trans. Roy. Soc. A255, 607, 633.

BATCHELOR, G.K. (1967) An introduction to fluid dynamics. C.U.P.

BATCHELOR, G.K. and GREEN, J.T. (1972) The determination of the bulk stress in a suspension of spherical particles to order c^2. J. Fluid Mech., 56, 401-427.

BLOCH, E.H. (1966) High speed cinephotography of the microvascular system. Proc. Fourth Europ. Conf. on Microcirc. Cambridge. Abstract F.I.

CARO, G.G., PEDLEY, T.J., SCHROTER, R.C. and SEED, W.A. (1978) The mechanics of the circulation. O.U.P.

COKELET, G.R. (1976) Macroscopic rheology and the tube flow of human blood. In Grayson, J. Singg, W. (Eds). "Microcirculation", Vol.1, New York. Plenum Press. pp. 9-31.

FUNG, Y.C. (1977) Rheology of blood in microvessels. In, Kaley, G. and Burton, M.A. (Eds). "Microcirculation", Vol.1, University Park Press, Baltimore, Chapter 13, pp. 279-298.

FUNG, Y.C. (1981) Biomechanics. Mechanical properties of living tissues. New York: Springer-Verlag, 1981.

GOLDSMITH, H.L. and KARINO, T. (1980) Physical and mathematical models of blood flow: Experimental studies. pp. 165-194. In: Erythrocyte mechanics and blood flow. Ed. Cokelet, G.R., Meiselman, H.J., Brooks, D.E. Publ. Alan R.Liss. Inc. New York.

HAPPEL, J. and BRENNER, H. (1965) Low Reynolds number hydrodynamics. Prentice-Hall.

KANG, C.K. and ERINGEN, A.C. (1976) The effect of microstructure on the rheological properties of blood. Bull. Math. Biol. 38, 135-159.

MEISELMAN, H.J. (1980) Measures of blood rheology and erythrocyte mechanics. pp. 75-117. In: Erythrocyte mechanics and blood flow. Ed. Cokelet, G.R., Meiselman, H.J., Brookes, D.E. Publ. Alan R.Liss. Inc. New York.

MOORE, W.J. (1962) Physical Chemistry. Longmans Green & Co.Ltd.

ROSENBLUM, W.I. (1977) Viscosity. 'In vitro' versus 'in vivo'. In, Kaley, G. and Burton, M.A. (Eds) "Microcirculation", Vol.1, University Park Press, Baltimore, Chapter 15, pp. 325-334.

THOMAS, H.W. (1962) The wall effect in capillary instruments: An improved analysis suitable for application to blood and other particulate suspensions. Biorheology 1: 41.

THOMAS, H.W., FRENCH, R.J., GROOM, A.C. and ROWLANDS, S. (1965) The flow of red cell suspensions through narrow tubes: the (extra corporeal) determination of the difference in mean velocities of red cells and their suspending phase. Proc. Fourth Int. Congr. on Rheol. 4, Symp. on Biorheol, e.g. Copley, A.L. Interscience, New York, pp. 381-391.

TROWBRIDGE, E.A. (1982a) The fluid mechanics of blood: Equilibrium and sedimentation. Clin. Phys. Physiol. Meas. (In press).

TROWBRIDGE, E.A. (1982b) The physics of arteriole blood flow. I. General Theory. Clin. Phys. Physiol. Meas. Submitted for publication.

TROWBRIDGE, E.A. and MEADOWCROFT, P.M. (1982) The physics of arteriole blood flow. II. Comparison with experiment. Clin. Phys. Physiol. Meas. Submitted for publication.

Models of oxygen transport in artificial oxygenators

A.R. VEITCH and G. EASON

SUMMARY

General equations are formulated to describe the oxygen transport into blood flowing through an artificial oxygenator. These equations are specialised to represent certain particular situations in duct or microtubular oxygenators. The trade-off between the detail represented and the mathematical tractability is explored.

1. INTRODUCTION

An artificial oxygenator is a device to replace the action of heart and lungs during heart surgery. In this paper and the sequel [1] attention is confined to that section of the oxygenator which replaces lung action.

Two designs are considered. In the first, oxygen depleted blood flows in a rectangular duct in which two wide parallel faces consist of membranes permeable to oxygen (fig. 1).

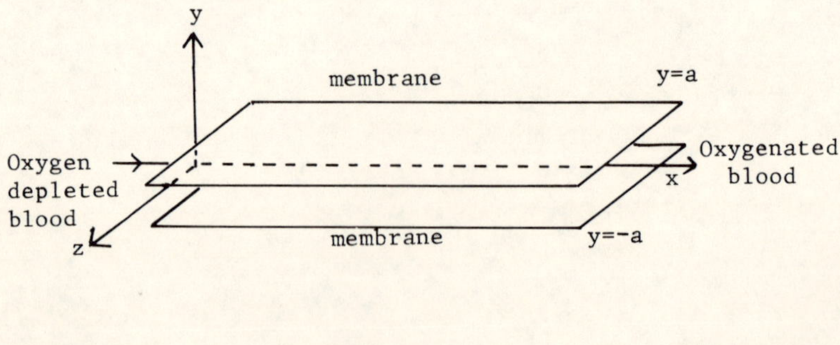

Fig. 1

An oxygen-rich atmosphere is maintained outside the membrane so that the blood becomes oxygenated as it passes along the duct.

In the second design blood flows through an arrangement of small-bore tubes made of a substance permeable to oxygen. These are enclosed in an outer container in which an oxygen-rich atmosphere is maintained (fig. 2).

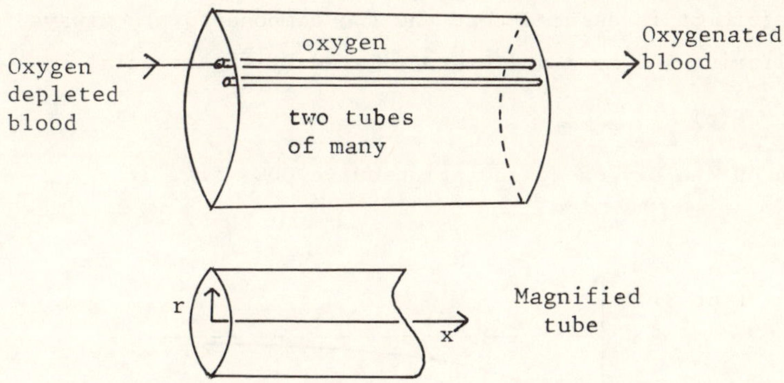

Fig. 2

In the turbulent flow oxygenator a typical separation of the membranes is 5mm and a typical length is about 500mm. For laminar flow membrane oxygenators the separation is much smaller and the device has multiple layers of alternating oxygen and blood separated by membranes. In the microtubular oxygenator the tube radii a are tiny, and so only laminar flow is of interest.

This chapter and its sequel deal with the mass-transfer of oxygen into the blood as it flows through the duct or through a typical tube. In §2 the basic equations are presented without reference to a particular design. Initial and boundary conditions are presented in general in §3, and then specialised to duct and to tube oxygenators. In §4 the equations are further simplified by introducing further reasonable assumptions, and in §5 the outcome of the whole programme of simplification is discussed. Some particular solutions are given in the accompanying paper.

2. BASIC EQUATIONS

Blood is taken to consist of two components, plasma and haemoglobin. The plasma holds dissolved oxygen with partial pressure p while the haemoglobin, by a chemical mechanism, holds a fraction σ of its maximum (saturation) capacity. Equilibrium between plasma and haemoglobin is achieved very quickly, in a time much less than the residence time of blood in the oxygenator. Therefore it is assumed that the two components are always in equilibrium and so σ and p are related by the formula

$$\sigma = f(p) \tag{2.1}$$

shown on the oxygen dissociation curve ODC (fig. 3).

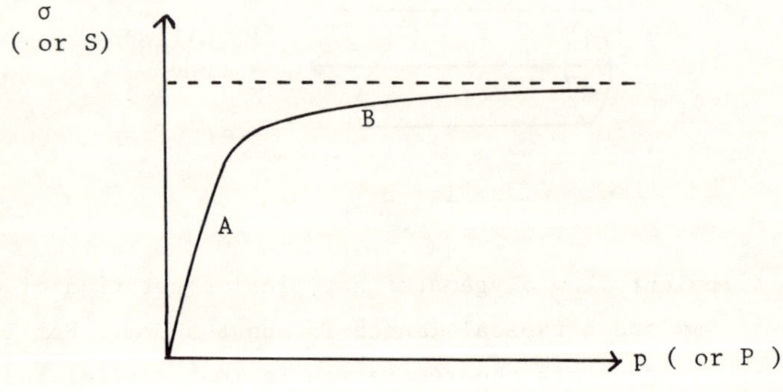

Fig. 3

Equation (2.1) is usually represented by the Margaria formula [2].

$$\sigma = \frac{\gamma p(1+\gamma p)^3 + \ell(\gamma p)^4}{(1+\gamma p)^4 + \ell(\gamma p)^4} \tag{2.2}$$

where γ and ℓ are constants. The essential feature is that, although sections A and B of the ODC separately are approximately linear, the whole ODC is distinctly nonlinear. It is unfortunate for the mathematical treatment that, in normal use, the oxygenator values of p and σ involve both sections.

Let P be the volume concentration of oxygen in the plasma, and S that in the haemoglobin. As P is proportional to p and S is proportional to σ (for a given level of haemoglobin in blood, a characteristic which may vary between patients) then the ODC can be re-expressed

$$S = F(P). \tag{2.3}$$

The graph of this is just fig. 3 with suitable scale changes on the axes.

In laminar flow oxygen is transported through the plasma by molecular diffusion with flux density $-D\nabla P$, where D is the diffusivity constant. So for *any* volume V of blood moving through an oxygenator, and not adjacent to a wall, the rate of loss of oxygen with time t is accounted for by the diffusion across the boundary ∂V:

$$-\frac{d}{dt}\int_V (P+S)\,dV = \int_{\partial V} -D\nabla P.\underline{dS}\ . \tag{2.4}$$

If the flow is turbulent the additional turbulent mixing can be represented as a further diffusive process involving both plasma and haemoglobin equally. The associated flux density is $-\varepsilon\nabla(P+S)$ where ε is the "eddy diffusivity" which varies with position depending on the flow conditions. Then (2.4) becomes

$$-\frac{d}{dt}\int_V (P+S)\,dV = -\int_{\partial V} D\nabla P.\underline{dS} - \int_{\partial V} \varepsilon\nabla(P+S).\underline{dS}\ . \tag{2.5}$$

In the usual way differentiation of the integral over a *moving* volume, use of the divergence theorem and the fact that V is *any* volume lead to the partial differential equation

$$\frac{\partial}{\partial t}(P+S) + \nabla.\{\underline{u}(P+S)\} = D\nabla^2 P + \nabla.\varepsilon\nabla(P+S), \tag{2.6}$$

where $\underline{u}$ is the blood velocity, taken to be given at each point in the flow. Equation (2.3) can be used to eliminate S in (2.6) leaving a differential equation in P. As it will be nonlinear this is not necessarily the best solution strategy; but it does show what initial and boundary conditions are needed.

3. INITIAL AND BOUNDARY CONDITIONS

The initial concentration is assumed to be given at each point of the oxygenator by

$$P = P_o \quad \text{at} \quad t = 0, \quad \text{all interior positions.} \tag{3.1}$$

Similarly at each point of the inlet surface

$$P = Q \quad \text{at} \quad x = 0, \quad \text{all } t > 0. \tag{3.2}$$

In the following the assumption is made that Q is constant across the inlet because of pre-mixing.

At the permeable boundaries the flux density of oxygen into the blood is taken to be $\alpha(R-P)$ where α is a constant mass transfer coefficient characteristic of the boundary, R is the oxygen concentration just outside the boundary and P the concentration just inside it. As this must balance the flux density into the blood the boundary condition becomes, in general,

$$D \frac{\partial P}{\partial n} + \varepsilon \frac{\partial (P+S)}{\partial n} = \alpha(R-P), \quad \text{on boundary}, \tag{3.3}$$

where $\frac{\partial}{\partial n}$ indicates the directional derivative in the direction of the normal $\underline{n}$ outwards from the blood. In the case where α is very large, so that penetration of the permeable wall is not the limiting feature in the oxygen transfer, equation (3.3) may be replaced by

$$P = R \quad \text{on the boundary}. \tag{3.4}$$

For duct geometry the effects of the side walls of the duct are neglected, and the duct is treated as a section of a flow between infinite parallel planes at y = a and y = -a. With ε taken as zero (at least on the boundary) (3.3) becomes

$$\pm \frac{\partial P}{\partial y} = \frac{\alpha}{D} (R-P) \quad \text{at} \quad y = \pm a. \tag{3.5a}$$

For tube geometry (3.3) becomes

$$\frac{\partial P}{\partial r} = \frac{\alpha}{D} (R-P) \quad \text{at} \quad r = a. \tag{3.5b}$$

A regularity condition at r = 0 is also required for this case, namely

$$P \text{ is bounded as } r \to 0. \tag{3.6b}$$

It is convenient for duct flows which are symmetric about y = 0 to consider only the solution in 0 < y < a. In this case

$$\frac{\partial P}{\partial y} = 0 \quad \text{at} \quad y = 0 \tag{3.6a}$$

replaces the condition at y = -a in (3.5a).

Finally a condition is needed at the outlet end of the

oxygenator. A detailed analysis has shown that a boundary layer (in concentration) exists to facilitate a rapid transition to any imposed downstream condition. These details are of little practical consequence; and what is really required is the solution appropriate to an oxygenator of infinite length, satisfying

$$P \text{ is bounded as } x \to \infty. \tag{3.7}$$

4. SPECIALISATIONS OF THE EQUATIONS

The case of turbulent flow in tubes has been treated numerically by Soda and Chow [3] for steady flow, and by Lim [4] for rectangular duct flow. In these cases specific detailed assumptions were made about $\underline{u}$ and ε in different regions of the flow to model the conditions as accurately as possible. Numerical methods provide the only way of treating the problem at this level of complexity. Great care is required with the discretisation because these convection-diffusion problems are notorious for the spread of truncation errors.

For the rest of this paper and its sequel laminar, unidirectional, incompressible flow is assumed with velocity U in the direction of the x-axis. Equation (2.6) then reduces to

$$\frac{\partial}{\partial t}(P+S) + U\frac{\partial}{\partial x}(P+S) = D\nabla^2 P \tag{4.1}$$

where for a duct

$$\nabla^2 P = \partial^2 P/\partial x^2 + \partial^2 P/\partial y^2 \tag{4.2a}$$

and for a tube, with radial coordinate r,

$$\nabla^2 P = \partial^2 P/\partial x^2 + \partial^2 P/\partial r^2 + (1/r)\partial P/\partial r. \tag{4.2b}$$

The best assumption for U is

$$U = U_o(1-y^2/a^2) \qquad \text{(duct)}, \tag{4.3a}$$

or

$$U = U_o(1-r^2/a^2) \qquad \text{(tube)}, \tag{4.3b}$$

where U_o is the centre-line velocity. A less accurate but more tractable assumption is of plug flow with velocity $\bar{U}_o$ found by (say) averaging so that

$$\bar{U} = \frac{2}{3} U_o \quad \text{(duct)} \quad \text{or} \quad \frac{1}{2} U_o \quad \text{(tube)}. \tag{4.4}$$

Analytical progress to date has depended upon the linearisation of (2.3), namely

$$S = kP + C. \tag{4.5}$$

This will overestimate S for larger values of P if linearised in section A of the ODC, and will overestimate S for smaller values of P if linearised in section B. With this linearisation a number of time-independent problems have been solved. Some of these are discussed in [1].

5. DISCUSSION

A programme of simplification has been outlined whereby equations of great generality have been simplified to treat particular situations. As is usual in modelling it is necessary to compromise between very detailed modelling assumptions and mathematical equations which can be handled analytically. It is not yet clear where this compromise is to be made.

Most of the simplifications are reasonable for practical purposes. The most serious one is the linearisation of the ODC which is not realistic for normal oxygenator operation. Further analytical progress depends upon perturbation methods, which are under study, or by piecing together solutions of linearised problems.

Numerical methods are needed to treat the problem in its full complexity. Published work has been centred on the Crank-Nicolson method for parabolic equations ($\partial^2 P/\partial x^2$ omitted). It is not yet clear whether the methods which produced the published results are adequate in speed and accuracy for all problems of the type discussed here.

REFERENCES

1. G. Eason, A.R. Veitch: "Some solutions of the basic equations governing oxygen transport in a membrane oxygenator". Following chapter.

2. R. Margaria: "A mathematical treatment of the blood dissociation curve for oxygen". Clin. Chem. 9 745-62 (1963).

3. K. Soda, J.C.F. Chow: "Blood oxygenation in tubular membrane oxygenator with turbulent flow". ASME paper 73-WA/Bio-23 (1973).

4. B-H. Lim: "Blood oxygenation in turbulent and laminar flow: macrochannel and microtubing configurations". Ph.D. thesis, University of Strathclyde, (1980).

Some solutions of the basic equations governing oxygen transport in a membrane oxygenator

G. EASON and A.R. VEITCH

SUMMARY

The basic differential equations which describe the transport of oxygen in blood in flat plate and tubular membrane oxygenators with laminar flow are summarised. Solutions are obtained for plug flow when the effect of axial diffusion is included in the basic equation and these solutions are compared with solutions obtained when this effect is neglected. The flat plate oxygenator with a parabolic velocity profile is also considered. The inclusion of axial diffusion leads to a solution in terms of non-orthogonal functions. This results in complications which have not been resolved.

1. INTRODUCTION

The basic differential equations governing the flow of blood in flat plate and membrane oxygenators have been reviewed by Veitch and Eason [1] and are simply stated here. The problem considered is that of a semi-infinite oxygenator, of either flat plate or tubular type, in which the flow is assumed to be laminar. This is a significant assumption since in some oxygenators a certain amount of turbulence is present. It is also assumed that a linear relationship exists between the plasma oxygen concentration and the oxygen concentration in the haemoglobin. This assumption simplifies the analysis considerably but places a restriction on the applicability of the results.

The principal concern of this paper is to examine the consequences of including the effect of axial diffusion in the basic equations. This effect has been neglected by previous authors. Its inclusion leads to equations which can be solved in certain cases. It is shown that for a typical membrane or microtubular oxygenator the omission of this term has a negligible effect on the solution. However, in one case the inclusion of axial diffusivity leads to a solution in terms of non-orthogonal functions and results in certain difficulties which have not been completely resolved.

In section 2 the basic equations for the various problems are summarised. The flat plate oxygenator with plug flow, with and without axial diffusion, is considered in section 3. Complete solutions in the form of series are obtained and compared. The similar problem for a tubular oxygenator with plug flow is considered in section 4. In section 5 the physically more important case of flow in a flat plate oxygenator with a parabolic velocity profile is considered. The solution in the absence of axial diffusion is known and is expressible in terms of orthogonal functions. When axial diffusion is included the resulting solution is in terms of non-orthogonal functions.

2. THE BASIC EQUATIONS

The problems to be considered here are all concerned with the laminar flow of blood in a flat plate or tubular oxygenator. The physical background of the problem has been discussed by Veitch and Eason [1] and by Soda and Chow [2] among others and no further discussion is given here. The basic equation governing the flow, which is equation (2.6) in Veitch and Eason, for incompressible, steady, non-turbulent flow reduces to

$$\underline{u} \cdot \nabla(P+S) = D\nabla^2 P, \tag{2.1}$$

where $\underline{u}$ is the velocity of the blood, P is the plasma oxygen concentration, S is the oxygen concentration in the haemoglobin and D is the effective diffusivity, assumed constant here. In addition it is assumed that a linear relationship (4.5) of Veitch and Eason [1] exists between S and P in place of the non-linear relationship (2.2) of [1] so that

$$S = kP + C, \tag{2.2}$$

where C and k are constants. This last assumption simplifies the analysis considerably but restricts the applicability of the solution to situations in which S and P do not vary too far from a constant value. Equations (2.1) and (2.2) may be combined to give the single equation

$$(1+k)\underline{u}.\nabla P = D\nabla^2 P. \tag{2.3}$$

Cartesian coordinates x,y,z are chosen so that the x-axis is in the direction of the blood flow. In straight-sided oxygenators the effects due to transport in the transverse direction may be neglected and (2.3) may be written

$$(1+k)U \frac{\partial P}{\partial x} = D\nabla^2 P, \tag{2.4}$$

where U is the velocity in the x-direction. A number of special cases of (2.4) are considered here and the particular forms are now stated.

For plug flow in a flat plate oxygenator in which the flow is independent of z

$$U = \bar{U}_o, \tag{2.5}$$

where $\bar{U}_o$ is a constant, and (2.4) reduces to

$$\frac{1}{K}\frac{\partial P}{\partial x} = \frac{\partial^2 P}{\partial x^2} + \frac{\partial^2 P}{\partial y^2}, \tag{2.6}$$

where

$$K = \frac{D}{(1+k)\bar{U}_o}. \tag{2.7}$$

When it is further assumed that the diffusive effect in the direction of flow is negligible then the first term on the right hand side of (2.6) may be neglected and (2.6) reduces to a diffusion equation.

In the case of plug flow in a tubular oxygenator (2.5) still applies and (2.4) gives

$$\frac{1}{K}\frac{\partial P}{\partial x} = \frac{\partial^2 P}{\partial x^2} + \frac{\partial^2 P}{\partial r^2} + \frac{1}{r}\frac{\partial P}{\partial r}, \tag{2.8}$$

where K is defined by (2.7) and r is the polar coordinate given here by

$$r = (y^2+z^2)^{\frac{1}{2}}. \tag{2.9}$$

A diffusion equation is again obtained when the first term on the right hand side of (2.8) is neglected.

A physically more realistic velocity profile is of parabolic form so that for the flat plate oxygenator with membranes at $y = \pm a$

$$U = U_o(1-y^2/a^2), \tag{2.10}$$

where U_o is constant. Equation (2.4) becomes

$$\frac{1}{K}\left(1 - \frac{y^2}{a^2}\right)\frac{\partial P}{\partial x} = \frac{\partial^2 P}{\partial x^2} + \frac{\partial^2 P}{\partial y^2}, \tag{2.11}$$

where K is again defined by (2.7) with $\bar{U}_o$ replaced by U_o. The equation for a tubular oxygenator with a parabolic velocity profile corresponding to (2.11) may similarly be written down using the right-hand side from (2.8).

Some particular solutions of equations (2.6), (2.8) and (2.11) will now be derived for an oxygenator of semi-infinite extent. The oxygenator is assumed to occupy $x \geq 0$ and either $-a \leq y \leq a$ or $r \leq a$ depending on the geometry. In all cases it is assumed that the appropriate velocity conditions occur at inlet ($x = 0$) and exist throughout the oxygenator. It is further assumed that

$$\begin{aligned} &\text{(i)} \quad P = Q \text{ on } x = 0, \\ &\text{(ii)} \quad P \text{ finite as } x \to \infty, \\ &\text{(iii)} \quad \frac{\partial P}{\partial y} = 0 \text{ on } y = 0 \text{ or } P \text{ finite on } r = 0, \\ &\text{(iv)} \quad P = R \text{ on } y = a \text{ or } r = a \text{ as appropriate,} \end{aligned} \tag{2.12}$$

where Q and R are constants. Condition (iv) is chosen for simplicity although as indicated by Veitch and Eason [1] other conditions may be more appropriate.

3. FLAT PLATE OXYGENATOR, PLUG FLOW.

(a) Axial Diffusion Neglected

When the axial diffusion term is neglected in (2.6) the resulting basic equation is the diffusion equation

$$\frac{1}{K}\frac{\partial P}{\partial x} = \frac{\partial^2 P}{\partial y^2} . \tag{3.1}$$

The solution of this equation that satisfies conditions (2.12) may be obtained using separation of variables and is found to be

$$P(x,y) = R + \frac{4}{\pi}(Q-R)\sum_{s=0}^{\infty}\frac{(-1)^s}{1+2s}\cos(ny/a)\exp(-Kxn^2/a^2), \tag{3.2}$$

where

$$n = \pi(1+2s)/2. \tag{3.3}$$

The normalised average value of the concentration P at any cross section is obtained from (3.2) in the form

$$T = \frac{1}{2a}\int_{-a}^{a}\frac{P-Q}{R-Q}\,dy$$

$$= 1 - \frac{8}{\pi^2}\sum_{s=0}^{\infty}\frac{1}{(1+2s)^2}\exp(-Kxn^2/a^2). \tag{3.4}$$

(b) Axial Diffusion Included

The inclusion of axial diffusion gives rise to equation (2.6). The solution of (2.6) that satisfies conditions (2.12) may be obtained by separation of variables in the usual way. It is found that

$$P(x,y) = R + \frac{4}{\pi}(Q-R)\sum_{s=0}^{\infty}\frac{(-1)^s}{1+2s}\cos(ny/a)\exp(-KxN^2/a^2), \tag{3.5}$$

where n is given by (3.3) and

$$N^2 = \frac{a^2}{2K^2}\{(1 + \frac{4n^2K^2}{a^2})^{\frac{1}{2}} - 1\}. \tag{3.6}$$

In this case the normalised average value of P is obtained by integrating (3.5) to obtain

$$T = 1 - \frac{8}{\pi^2}\sum_{s=0}^{\infty}\frac{1}{(1+2s)^2}\exp(-KxN^2/a^2). \tag{3.7}$$

It is of interest to compare the solution obtained by including axial diffusion (3.5) with that obtained by neglecting this term. This comparison may be made by assuming that $D/a\bar{U}_o$ is small in (2.7) so that K/a is taken to be small in (3.6). An expansion of (3.6) for small K/a gives $N^2 \approx n^2$ as is to be expected. This expansion is valid provided that n is not large. However, due to the exponential decay in x present in (3.2) and (3.5) contributions from terms involving large values of n will not be significant except, possibly, when x is small. Thus the solution obtained by neglecting axial diffusion, (3.2), is likely to give good results in this case.

This view is upheld by the following table of normalised average concentrations at various cross-sectional positions defined by the dimensionless coordinate Kx/a^2. As discussed above $K/a \to 0$ corresponds to the solution (3.4) and the other columns refer to solution (3.7) for various values of K/a.

Kx/a^2	$K/a \to 0$	K/a=0.1	K/a=0.5	K/a=1.0
0.01	0.113	0.097	0.052	0.033
0.02	0.160	0.146	0.087	0.058
0.05	0.252	0.242	0.167	0.116
0.10	0.357	0.349	0.264	0.192
0.20	0.504	0.497	0.404	0.307
0.50	0.764	0.757	0.656	0.532
1.00	0.931	0.927	0.855	0.742
2.00	0.994	0.993	0.974	0.918
5.00	1.000	1.000	1.000	0.997

A typical value of K/a for a multilayer membrane oxygenator is about 0.0001. For this value the normalised average concentrations are indistinguishable from those for $K/a \to 0$.

4. TUBULAR OXYGENATOR, PLUG FLOW.

(a) Axial Diffusion Neglected

In the absence of axial diffusion (2.8) becomes

$$\frac{1}{K}\frac{\partial P}{\partial x} = \frac{\partial^2 P}{\partial r^2} + \frac{1}{r}\frac{\partial P}{\partial r}, \tag{4.1}$$

and the solution satisfying conditions (2.12) is

$$P(x,r) = R + 2(Q-R) \sum_{s=1}^{\infty} \frac{J_o(\lambda_s r/a)}{\lambda_s J_1(\lambda_s)} \exp\{-Kx\lambda_s^2/a^2\}, \tag{4.2}$$

where $J_o(\mu)$ is the Bessel function of the first kind of zero order and the λ_s are roots of

$$J_o(\lambda) = 0. \tag{4.3}$$

(b) Axial Diffusion Included

Equation (2.8) is obtained when axial diffusion is included. The solution of this equation is straightforward using the method of separation of variables. Applying conditions (2.12) it results in

$$P(x,r) = R + 2(Q-R) \sum_{s=1}^{\infty} \frac{J_o(\lambda_s r/a)}{\lambda_s J_1(\lambda_s)} \exp(-KxN^2/a^2), \tag{4.4}$$

where N^2 is given by (3.6) with n replaced by λ_s. Comparison of (4.2) with (4.4) leads to conclusions identical with those arrived at in section 3 for the flat plate. Again it is to be expected that (4.2) will give satisfactory results in this case.

5. FLAT PLATE OXYGENATOR, PARABOLIC PROFILE

The parabolic velocity profile (2.10) gives a more satisfactory description of the flow of a viscous fluid such as blood than does plug flow. For this reason it is of interest to determine the effect of neglecting axial diffusion for this flow. As will be seen the comparison is not as simple as in the case of plug flow.

(a) Axial Diffusion Neglected

The neglect of axial diffusion in (2.11) results in the equation

$$\frac{1}{K}\left(1 - \frac{y^2}{a^2}\right)\frac{\partial P}{\partial x} = \frac{\partial^2 P}{\partial y^2} \quad . \tag{5.1}$$

Solutions of this equation have been presented by Walker and Davies [3] among others. Solutions may be obtained by separation of variables or by assuming that

$$P_n(x,Y) = g_n(Y) \exp(-n^2Kx/a^2), \tag{5.2}$$

where n is a constant, as yet undetermined,

$$Y = y/a, \tag{5.3}$$

and $g_n(Y)$ is a function of Y alone. The equation satisfied by g_n is

$$\frac{d^2 g_n}{dY^2} + n^2(1-Y^2)g_n = 0, \tag{5.4}$$

and the solution of this equation that satisfies the third of conditions (2.12) is

$$g_n(Y) = A_n \, {}_1F_1(\tfrac{1}{4}(1-n);\ \tfrac{1}{2};\ nY^2)\ \exp(-\tfrac{1}{2}\,nY^2), \tag{5.5}$$

where A_n is a constant and ${}_1F_1(\alpha;\rho;\eta)$ is the confluent hypergeometric function defined by

$${}_1F_1(\alpha;\rho;\eta) = 1 + \frac{\alpha\eta}{1!\rho} + \frac{\alpha(\alpha+1)\eta^2}{2!\rho(\rho+1)} + \frac{\alpha(\alpha+1)(\alpha+2)\eta^3}{3!\rho(\rho+1)(\rho+2)} + \ldots\ . \tag{5.6}$$

The remainder of conditions (2.12) must be satisfied in turn. It is found that appropriate values of n are $n = n_s$ determined by roots of

$${}_1F_1(\tfrac{1}{4}(1-n);\ \tfrac{1}{2};\ n) = 0. \tag{5.7}$$

The final solution is

$$P(x,y) = R + (Q-R)\sum_{s=1}^{\infty} B_s\phi_s(Y)\ \exp(-Kxn_s^2/a^2), \tag{5.8}$$

where

$$\phi_s(Y) = {}_1F_1(\tfrac{1}{4}(1-n_s);\ \tfrac{1}{2};\ n_sY^2)\ \exp(-\tfrac{1}{2}\,n_sY^2). \tag{5.9}$$

It may be shown that

$$\int_0^1 (1-Y^2)\ \phi_s(Y)\ \phi_t(Y)dY = 0, \quad s \neq t, \tag{5.10}$$

and B_s is determined by

$$B_s\int_0^1 (1-Y^2)\ \phi_s(Y)\ \phi_s(Y)dY = \int_0^1 (1-Y^2)\phi_s(Y)dY. \tag{5.11}$$

Thus, in the absence of axial diffusion, the solution is obtained in terms of functions $\phi_s(Y)$ which are orthogonal and the problem is a typical Sturm-Liouville problem.

It is of interest to note that the solution for $g_n(Y)$ may also be expressed in terms of parabolic cylinder functions or Whittaker functions.

(b) Axial Diffusion Included

When the axial diffusion term is included it is necessary to solve (2.11). When a solution of the form

$$P_N(x,Y) = G_N(Y)\exp(-KxN^2/a^2), \tag{5.12}$$

similar to (5.2), is assumed it is found that G_N satisfies the equation

$$\frac{d^2G_N}{dY^2} + N^2\{\frac{N^2K^2}{a^2} + (1-Y^2)\}\, G_N = 0. \tag{5.13}$$

The solution of this equation that satisfies the third of conditions (2.12) is

$$G_N(Y) = A'_N\, {}_1F_1(\alpha_N;\, \tfrac{1}{2};\, NY^2)\exp(-\tfrac{1}{2}NY^2), \tag{5.14}$$

where A'_N is constant and

$$\alpha_N = \tfrac{1}{4}(1 - N - N^3K^2/a^2). \tag{5.15}$$

The remainder of conditions (2.12) must now be satisfied. The fourth condition gives the equation

$${}_1F_1(\alpha_N;\, \tfrac{1}{2};\, N) = 0, \tag{5.16}$$

from which to determine suitable values of $N = N_s$. The solution may now be written in the form

$$P(x,Y) = R + (Q-R)\sum_{s=1}^{\infty} B'_s\,\psi_s(Y)\exp(-KxN_s^2/a^2), \tag{5.17}$$

where B'_s are constants,

$$\psi_s(Y) = {}_1F_1(\beta_s;\, \tfrac{1}{2};\, N_sY^2)\exp(-\tfrac{1}{2}N_sY^2), \tag{5.18}$$

and

$$\beta_s = \tfrac{1}{4}(1-N_s-N_s^3\, K^2/a^2). \tag{5.19}$$

The first of conditions (2.12) with (5.17) gives

$$\sum_{s=1}^{\infty} B'_s \psi_s(Y) = 1, \tag{5.20}$$

as the basic equation from which to determine the values of the coefficients B'_s.

The functions ψ_s satisfy the differential equation

$$\frac{d^2\psi_s}{dY^2} + N_s^2 \{N_s^2 \frac{K^2}{a^2} + (1-Y^2)\}\psi_s = 0. \tag{5.21}$$

When the usual orthogonality procedure is applied to (5.21) it is found that

$$\int_0^1 (1-Y^2)\ \psi_s(Y)\ \psi_t(Y)dY$$
$$= -\frac{K^2}{a^2}(N_s^2+N_t^2)\int_0^1 \psi_s(Y)\ \psi_t(Y)dY,\ s \neq t, \tag{5.22}$$

so that the ψ_s are not orthogonal and it is not possible to determine the constants B'_s from (5.20) in the usual way. Multiplying (5.20) by $(1-Y^2)\ \psi_t(Y)$ and integrating from 0 to 1 with respect to Y results in the infinite set of equations

$$\sum_{s=1}^{\infty} B'_s J_{st} = I_t, \tag{5.23}$$

where

$$I_t = \int_0^1 (1-Y^2)\ \psi_t(Y)dY, \tag{5.24}$$

$$J_{st} = \int_0^1 (1-Y^2)\ \psi_s(Y)\ \psi_t(Y)dY. \tag{5.25}$$

The result (5.22) may be applied to (5.25) for $s \neq t$. The calculation of B'_s from (5.23) requires the inversion of an infinite symmetric matrix in which the terms in the leading diagonal will normally dominate those off the leading diagonal due to (5.22).

In order to determine the B'_s some process of approximation must be applied such as truncation of the equations or the application of the Gram-Schmidt process. In the present case the more laborious Gram-Schmidt process appears to be equivalent to truncation up to the constant B'_3 and possesses no advantages.

For suitably small values of K/a it is clear that β_s given by (5.19) is approximately $\frac{1}{4}(1-N_s)$ which corresponds with the value $\frac{1}{4}(1-n_s)$ appearing in (5.9). It is also evident that when K/a is small the right hand side of (5.22) approaches zero and the ψ_s are almost orthogonal.

6. CONCLUSION

Solutions of the basic differential equations governing the behaviour of flat plate and tubular oxygenators have been presented for plug flow and parabolic velocity profiles. When the effect of axial diffusion is included it is shown that the solution for plug flow may be obtained in a straightforward manner. However for the parabolic velocity profile the solution obtained is not in terms of orthogonal functions and this leads to complications which have not as yet been resolved.

REFERENCES

1. A.R. Veitch and G. Eason: Models of oxygen transport in artificial oxygenators, preceding chapter.
2. K. Soda and J.C.F. Chow: Blood oxygenation in tubular membrane oxygenator with turbulent flow, ASME paper 73 - WA/Bio - 23, American Society of Mechanical Engineers Winter Annual Meeting, Detroit, 1973.
3. G. Walker and T. Davies: Mass transfer in laminar flow between parallel permeable walls, A.I. Ch. E. J., 20, 881(1974).